THE INNER WORKINGS OF LIFE

Living systems are dynamic and extremely complex and their behavior is often hard to predict by studying their individual parts. Systems biology promises to reveal and analyze these highly connected, regulated, and adaptable systems, using mathematical modeling and computational analysis. This new systems approach is already having a broad impact on biological research and has potentially far-reaching implications for our understanding of life. Written in an informal and non-technical style, this book provides an accessible introduction to systems biology. Self-contained vignettes each convey a key theme and are intended to enlighten, provoke, and interest readers of different academic disciplines, but also to offer new insight to those working in the field. With subtle wit and eloquence, Voit manages to convey complex ideas and give the reader a genuine sense of the excitement that systems biology brings with it, as well as the current challenges and opportunities.

∼

Eberhard O. Voit is a pioneer and leader in systems biology, with a passion for education at all levels. He is Professor and David D. Flanagan Chair in Biological Systems, as well as a Georgia Research Alliance Eminent Scholar in the Wallace H. Coulter Department of Biomedical Engineering at the Georgia Institute of Technology and Emory University. He is the author of one of the leading textbooks in the field, wrote one of the first systems biology books published, and has written over 250 scientific articles. His research focuses on genomic, metabolic, and signaling systems, with applications reaching from microbial, ecological, and plant systems to human diseases.

T0188338

THE INNER WORKINGS OF LIFE
VIGNETTES IN SYSTEMS BIOLOGY

~

Eberhard O. Voit

Georgia Institute of Technology, USA

CAMBRIDGE
UNIVERSITY PRESS

University Printing House, Cambridge CB2 8BS, United Kingdom

Cambridge University Press is part of the University of Cambridge.

It furthers the University's mission by disseminating knowledge in the pursuit of
education, learning and research at the highest international levels of excellence.

www.cambridge.org
Information on this title: www.cambridge.org/9781316604427

© E.O. Voit 2016

First published 2016

Printed in the United Kingdom by Clays, St Ives plc

A catalogue record for this publication is available from the British Library

Library of Congress Cataloguing in Publication data
Names: Voit, Eberhard O., author.
Title: The inner workings of life : vignettes in systems biology / Eberhard O. Voit.
Description: Cambridge; New York : Cambridge University Press, 2016. |
Includes bibliographical references and index.
Identifiers: LCCN 2016005715| ISBN 9781107149953 (hardback) |
Subjects: | MESH: Systems Biology
Classification: LCC QH324.2 | NLM QU 26.5 | DDC 570.1/13–dc23
LC record available at http://lccn.loc.gov/2016005715

ISBN 978-1-107-14995-3 Hardback
ISBN 978-1-316-60442-7 Paperback

Contents

Contents

Appetizer

Yes, I called this first section *Appetizer* because *Preface* or *Foreword* is just too boring. Many a preface consists of a laundry list of topics, which the author felt obliged to include, followed by acknowledgments in often cumbersome prose. And even if a preface actually contains true morsels of wisdom, readers don't expect as much and don't do what readers are supposed to do: read it. An appetizer, by contrast, is an artfully created tidbit to whet your appetite. Here it goes loosely with the theme of vignettes, that is, wine, vine leaves and stories that are short enough to fit on them. It is also more in line with the somewhat provocative chapter titles such as *I'd rather be fishin'*, but I will not launch into a laundry list. Mind you, esteemed reader, you are still with me, and I hope you will keep on reading until, as your mother might have said, you have earned the privilege to enjoy the *Dessert*. In the process, I hope to ease you into the fascinating field of systems biology as it attempts to decipher and understand the inner workings of life. As you will see, systems biology uses a lot of sophisticated techniques, all with their own jargon, so this is not exactly an easy topic to discuss. But there are so many exciting things going on in this young field, and in biology in general, that I am convinced you would be at a loss if you remained in the dark. You did not have the opportunity to experience the industrial revolution firsthand, and you may have missed the electronics revolution, so don't miss out on the biological revolution!

Allow me to back up a little and provide some context for the rise of systems biology. To some degree, we all have been systems biologists since our early days. We have marveled at ecological systems in our backyards and across the world, in deserts, on mountains, and in oceans, either with our own eyes and ears, or maybe through magazines and documentaries. We have watched with amazement how tiny ants carry heavy loads into their nests, apparently knowing that this is their assigned task in life and that other ants have their own roles and responsibilities for the greater good of the colony. Time and again we have been perplexed by the beauty of butterflies siphoning nectar from flowers, and the intricacies of complicated food webs that have evolved over eons and have sustained and renewed themselves from one year to the next, in spite of – or maybe due to – their reliance on dozens of plant and animal species.

When we literally dig deeper and analyze a tablespoon of soil, we find not only plant roots and worms and insects, but also hundreds if not thousands of microbial species. They form complex and constantly adapting systems of their own, and without them plants and animals simply could not thrive. Similarly, our gut, our skin, our oral cavity are all home to thousands of microbial species, most of which are beneficial and indeed necessary for our health. But the complexity does not stop there. Each microbe, each plant or animal cell contains myriad genes, proteins, sugars, lipids and uncounted other molecules that play their specific roles for the organism's inner workings and are collectively essential to the well-being of the community of life.

The more we learn about these molecular systems, the more even hard-core scientists are in awe of the enormous complexity with which the components in these systems interact as self-organizing and controlling machines, reacting to signals from the outside and collaborating to mount appropriate and well-coordinated responses. These sub-microscopic systems are so robust and tolerant to all kinds of disturbances that we often take the living world around us for granted. Only if something goes wrong, if our well-being is compromised, whether the problem is a headache, an infection, or the onset of neurodegenerative disease, do we realize how intricate and vulnerable these systems are, and how processes, often entirely unknown to us before, can derail our existence when they fail. Intriguingly, the molecular systems in our cells involve thousands of different molecules and processes, yet they work together so smoothly, and usually adapt so effectively to altered conditions, that we don't even notice their normal function or their compensatory activities and internal rearrangements unless something is out of order.

Systems biology has declared that its goal is to elucidate and understand the functioning and regulation of biological systems under normal and disturbed conditions. That is a tall order, but then again, the potential of reaching, or even just approaching, the goal promises essentially unlimited rewards. Just imagine how well we could address cancer in a personalized manner if we really understood what had gone wrong. Try to fathom the possibilities of reliably steering microorganisms or cells into making valuable medicines cheaply or into churning out lots of food products or energy without negative side effects.

The systems of life contain many different "moving parts," and it is therefore no surprise that their exploration requires a rich mix of diverse expertise. Clearly, one needs vast amounts of data, which immediately implies many types of biomedical investigation, but also the necessity of computer support for data handling and analysis. In addition, systems biology draws heavily from chemistry and physics. It needs sophisticated pure math to decode systems and formulate the laws that govern them, as well as applied math, on which all computer algorithms are based. Last but not least, the field could not possibly prosper without a constant stream of ingenious engineering feats that enable, automate, and miniaturize experimental explorations of ever finer biological details.

As a novel concoction of biology, chemistry, physics, math, computer science and engineering, systems biology is as yet a largely untested, but very promising, scientific endeavor. It has received quite a bit of buzz in recent years, to a point where it has become fashionable to sneak "systems biology" into the titles of articles or applications for research funding. Yet, outside the trenches, many laypeople and scientists alike only have a vague idea of how exactly systems biology intends to attack some of the biomedical grand challenges of our times, what one might realistically expect to be achievable, and what the current obstacles and limitations of systems biology are.

Several textbooks on systems biology have entered the market in recent years, but by the nature of the beast, these publications rely heavily on biological and mathematical jargon. This bridging between actual biological entities and the abstract language of mathematics is necessary to advance the field, and it can be very beautiful – to the expert. Unfortunately, it also triggers in some the all too familiar knot in the stomach that is often associated with everything mathematical. As a result, the curious and well-intended, yet unsuspecting, casual browser in a bookstore may be intrigued by the idea of trying to find out how life works but, opening any current systems biology book, finds so many unfamiliar or long-forgotten symbols and equations, and so much jargon of molecular biology, that the only way to save face seems to be a stroll to an aisle in the bookstore that offers something lighter, much lighter.

I started pondering the feasibility of a book on *Systems Biology without Equal … Signs* many years ago. At first, it was difficult to envision doing the new field of systems biology justice without discussing computing and math, but I convinced myself that there is a big difference between making music or movies and sitting in the audience listening or watching. In fact, deep knowledge of all that's behind a literary play or a piece of art sometimes diminishes the visceral enjoyment that non-experts experience. Most of us do not have the opportunity, or are simply not fit enough, to climb Mount Kilimanjaro, and we will never feel the exhilaration of having made it to the top. Yet, we can vicariously enjoy the views through travelogs on television or in an IMAX theater. Borrowing from these analogies, I hope that you, the reader, will somewhat vicariously enjoy witnessing how we systems biologists are beginning to explore and decipher and understand the inner workings of life.

This book is certainly not an 8:00 a.m., hard-core text that is to be studied for an advanced college class. It is much too casual and non-technical for that. Rather, it is a curling-up-in-a-hammock-with-milk-and-cookies-in-a-summer-breeze read or maybe a deep-thoughts-about-the-secrets-of-life-at-midnight-with-scotch-and-candlelight treatise. It consists of vignettes, each of which is only a few pages long. These vignettes are presented in a loosely logical order, but they are all self-contained. My hope is that each vignette conveys a key thought that enlightens, provokes, interests, is maybe even fun to read, and leads to lasting thoughts that could easily float beyond biology and enter the philosophical realm of life.

Acknowledgments

Notwithstanding the non-preface, I am truly thankful for the support, curiosity, and intellectual stimulation that I regularly receive from my family, students, and peers. Most of all, I really do want to thank Ann Voit and James Wade for reading the entire draft manuscript and giving me uncounted valuable suggestions for improvements. I also express my appreciation to Drs. Edward Harpham, Manu Platt, Rabin Tirouvanziam, and three reviewers, who gave me excellent feedback on some of the material, Vickie Okrzesik for the beautiful cover, as well as Roland Voit for his help figuring out how many notes are in a Beethoven symphony; yes, stay tuned! Finally, I express my gratitude to Katrina Halliday and Victoria Parrin at Cambridge for their most effective and supportive shepherding of the book through the process, from my initial inquiry all the way to its run off the printing press.

1

Status: it's complicated!

In 1635, the English scientist Robert Hooke made a fantastic discovery. Studying a slice of cork through a microscope, he discovered cavities that, as he said, used to contain the "noble juices" that once had nurtured the living tree from which the cork had been cut. He called the cavities *cellulae,* a Latin word for storage rooms and the root of the term that we still use today: cells. Because the cork was dead, Hooke was only able to see cell walls forming a honeycomb structure. As exciting as this discovery must have been at the time, we now know that Hooke merely saw the tip of the iceberg, actually missing most of what makes a *cellula* such an impressive object. Peeking into a cell today with an optical or electron microscope, we see how a whole new world of structures and molecules opens up. Most cells have a nucleus, mitochondria, and ribosomes, and there are all kinds of small organelles, vesicles, and membrane enclosures. Going even further, modern visualization and tagging techniques of molecular biology allow us to see more and finer structures, all the way down to the level of large individual molecules. A whole world, invisible to the human eye, is emerging, leaving no doubt: life is complicated!

We systems biologists love to work on real puzzles. For many of us, living systems are huge Sudokus, where some information is available, but lots of gaps in-between are to be filled in through experiments and with advanced logic that evaluates the experimental results in a systemic context. Trying to figure out how the multitudinous parts in cells work together to create something as incredible as a brain is very attractive to us. We are fully aware that we will not solve the whole puzzle in our lifetimes, but nature is modular, and every systems biologist hopes to solve a large sub-puzzle, or at least a few smaller puzzles. The intellectual challenge is the enormous complexity of every cell and organism, which requires us to invent new tools and methods, and that's what systems biology is all about.

The complexity of living systems is due to different features. First and foremost, there are just very many parts. The lowly bacterium *E. coli* contains between four and five thousand genes. Nobody is sure how many different proteins are in a single plant or animal cell. Suffice it to say, there are easily tens of thousands. Our brains allegedly contain several hundred trillion connections between neurons, and the human body altogether supposedly consists of roughly five octillion atoms, that is:

5,000,000,000,000,000,000,000,000,000.

Now, it is unlikely that we will ever have the need to follow each atom explicitly, but it does seem rather evident that a deep understanding of the inner workings of life will require at least a good account of our genes and proteins. And therein lies our first grand challenge. The cellular Sudoku consists of a huge number of grid boxes. These sheer numbers already tell us that we need to generate our own support strategies, starting with tools that can handle the enormous bookkeeping tasks we have at our hands. Fortunately, computers are very good at dealing with gazillions of data points and not forgetting any of them.

While storing information is a considerable challenge, data points by themselves are usually not all that stimulating. Much more intriguing are the interwoven processes that lead to the data and, in particular, features of biological systems that look quite harmless at first, but can really play games with our minds. One such feature that we often take for granted, but that biological systems often violate, is linear scaling. We like to expect that there is a strict correlation between an input and the corresponding output. If we invest $1,000 in the stock market and after a couple of years receive $1,100 in return, we would have received $110,000 had we invested $100,000. Not so in biology. If we fertilize our roses with 2 tablespoons of fertilizer, they might produce 50 blossoms, but fertilizing them with 200 tablespoons will most assuredly not result in 5,000 blossoms. Rather, the roses will probably die. Thus, more is not necessarily better in biology, and there is often a strongly reduced return on investment. Many biological phenomena are even more complicated in that both very large and very small inputs are disadvantageous, if not lethal, while normal functionality requires an input of just the right intermediate magnitude. Other systems, particularly in the microbial world, are so robust that they can tolerate quite large variations in their environments without changing their functionality in response. So, the assumption of linearity is more often than not a problem in biology, and it is clear that we need nonlinear thinking. But that is a real challenge as nonlinear input–output relationships make it difficult for us to predict responses to perturbations to which a cell or organism is exposed.

Related to the issue of linear scaling is the principle of superposition, which is a cornerstone of many applications in engineering and in daily life. It addresses the relationship between inputs to a machine or system and the outputs with which it responds. The principle says that, if some Stimulus-1 leads to Response-1 and if Stimulus-2 leads to Response-2, then the Response to Stimulus-1 and Stimulus-2 together is equal to the sum of Response-1 and Response-2. As an example, consider a car on a suspension bridge. Due to the car's weight the bridge bends down very slightly. For a truck, the bending is a little stronger. If the car and the truck are both on the bridge, the total bending is the sum of the two. A typical example from physics is the force (vector) that results from two independent forces (vectors).

Biological systems often operate quite differently. Two inputs may lead to a much stronger response than expected or, in other cases, the response might be much weaker. More than 2,000 years ago, Aristotle already pondered what has almost become a cliché; namely, that a system can be more than the sum of its parts and that there may be synergism between actions or processes. He did not make up the fancy word *synergism* for this observation, as it is simply Greek for *collaboration*. Instead of collaborating and enhancing each other, it is also possible that processes work against each other and, in an antagonistic manner, diminish each other's impact. As an example, suppose that a cell can generate a metabolite through two pathways. If the activity of one is increased, the activity of the other is usually slowed down. An important advantage provided by the superposition principle, if it applies, is that it is legitimate to analyze parts of a system with respect to their input–output relationships one at a time. Recording all results then allows valid predictions of how the system will respond to combined inputs. Many engineered systems are designed such that the superposition principle holds and this type of analysis can be performed. The situation is dramatically different in a synergistic biological system, which often shows responses that none of the components by themselves can explain and where parts or subsystems, taken out of their contexts, frequently cease to function altogether. A lot of research has been devoted to creating artificial environments in which subsystems work properly, but the fact that isolated parts act differently if they are inside or outside their normal milieu remains a formidable challenge.

Western tradition trains us to think in chains of causes and effects. Our brain is good at following these chains, even if they are as convoluted as a Rube Goldberg contraption. However, as soon as the situation is a little bit more complicated, we scratch our heads. Let's do a thought experiment with the ubiquitous example of a negative feedback loop. Imagine a chain of events involving a gene G, which, when expressed, leads to the formation of the matching mRNA R, which is translated into a protein P. Let's assume that P is an enzyme that catalyzes the production of metabolite M. If that is all there is, then we easily predict what happens if G is expressed. Namely, in strict order we will see R, P, and finally M appear or rise in amount. If G is turned off, R, P, and M will eventually disappear. Now suppose a seemingly small addition to the system in the form of M directly affecting the expression of G. This situation is quite frequent and we find it, for instance, in the famous lactose genes that led to an enormous spectrum of insights into gene regulation and for which François Jacob, André Lwoff, and Jacques Monod obtained a Nobel Prize. Let's suppose that M represses the expression of G. Our natural thought process then is probably the following. If G is expressed, we find more R, more P, and more M. Now, the increase in M feeds back and represses G. As a consequence, we'll have less G, less R, less P, and less M. Less M means less repression of the expression of G, which should lead to more G, R, P, M, and so on. What should we conclude? The system appears to oscillate, but does it really and, if so, for how long? Hmm. We might believe there must be a way of solving the puzzle by thinking harder, but the truth is that we cannot figure it out with the information given. The reason is that the response of such a feedback system depends critically on the numerical features of the system, such as the number of events in the chain, the time delay between the expression of G and the production of M, and the strength of the feedback. As a daily life example, imagine a picky and impatient copilot in your car who finds the temperature much too cold. So he turns on the heater. It takes a while for the car to warm up, and all of a sudden it is too hot. So he turns the heating off. If the temperature adjustments are done in a well-measured manner, the temperature will eventually be just right, but if the copilot keeps overreacting with his feedback, it may oscillate between too cold and too hot for a long time. Nature is full of feedbacks, and they are often nested in complicated patterns. Biological systems also use feedback activation, as well as feedforward regulation. Furthermore, signals may compete with each other. If a process is simultaneously activated by one factor and inhibited by another, which one will win? Actual systems, even of moderate size, contain many such controls, and these often form complex regulatory webs. They constitute intriguing puzzles, which we can only solve effectively by setting up mathematical models.

The regulatory webs in cells and organisms often work simultaneously on different organizational scales and different time scales. For instance, cells exposed to physical or chemical stresses may alter their gene expression and protein profile, as well as their metabolite concentrations, in response. Multicellular organisms furthermore show different fast and slow physiological responses, such as shivering for immediate warming and creating a fat layer to ward against cold temperatures that occur on a regular basis. Over short periods, regulatory control systems ensure that the cell or organism remains close to its normal state of homeostasis. However, if the system is perturbed for a long time, the regulatory web mounts an adaptive, long-lasting response. In many cases this adaptation is successful, and the cell or organism lives essentially a normal life, even though it is exposed to inferior conditions. In other cases, the system may settle in a different, suboptimal state, such as a disease.

A good example of slow and fast adaptation is the reward system in the human brain. This system responds to pulses of neurotransmitters, such as dopamine and glutamate. Eating a piece of chocolate or looking at a beautiful sight triggers dopamine production and makes us feel good. However, if we repeatedly trigger or even overwhelm the reward system, for instance with the recreational drug methamphetamine, the number of dopamine receptors eventually begins to change, and we require ever-more input to achieve the same reward of feeling good. This regulatory response occurs at several organizational levels in the brain and involves electrical signals, metabolites, proteins, genes, and the restructuring of cell membranes.

A different, deceivingly simple-looking puzzle comes from threshold effects: a slight increase in some concentration or amount leads to a correspondingly slight change in some output, whereas a stronger increase has a totally different effect. An intuitive illustration is a visit to the beach. If we stay for a short period of time, our skin does not respond much; if we stay longer, we may develop a tan; if we stay too long, we go home with a bad sunburn. Tanning and sunburn are very different biological responses, but unless we have a lot of experience with our own body, it is difficult to predict when exactly one is replaced by the other. We experience the same type of input (sunlight) in both cases, but somewhere there is a threshold in its amount, which distinguishes one response from the other. Thresholds are very common in nature, and one might speculate that many diseases are caused by processes exceeding their "normal" thresholds. The challenge is that we do not really know what these thresholds are in most cases, and if several thresholds are in play, predictions regarding the responses of the system become very difficult. Moreover, thresholds frequently change over time. Systems get used to stresses, they adapt, and this adaptation may happen within hours, months, or on an evolutionary time scale.

Our innate way of linear thinking in terms of causes and effects has dominated biological research for a long time. In particular, it is directly in line with the paradigm of reductionism. The core idea of reductionism is that knowledge of all parts of a biological machine will tell us how the machine works. Therefore, understanding how an organism functions requires that we understand what its organs do. To understand organs, we investigate tissues and cells. In order to understand cells and their function, we study the details of intracellular structures, processes, and molecules. The implicit expectation is the following: if we work hard enough and characterize every constituent within a cell, we will grasp the secrets of biology. There is no doubt that we need to know the parts of biological systems and their features, but is this knowledge sufficient? The answer is no; we need additional techniques for putting the system together again.

A prominent example is the Central Dogma, which Nobel Laureate Francis Crick proposed about half a century ago and which somewhat simplistically states the following. Genes consist of DNA. DNA is transcribed into a matching RNA, and this RNA is translated into a string of amino acids, that is, a peptide or protein. Proteins are responsible for the processes of daily life; for instance, by providing physical structure and serving as signaling molecules or enzymes that control metabolism. The Central Dogma proposes a neat linear chain of causes and effects, and the entire blueprint of life therefore appears to be coded within the genes. Following through with this argument, it seems that we need to know the genes, and all else will follow. The reductionist paradigm therefore emphasizes the identification of genes and genomes very strongly.

The basic tenets of the Central Dogma are still undisputed, and genes do contain an enormous amount of information. However, the more we learn, the more we fathom how complicated the details are. We now know that the process of expressing a gene is controlled by transcription factors, which are proteins, and also by the three-dimensional structure of DNA and the way it is stored in cells. Additionally, expression may be affected by repressors or inducers, for instance in the form of the metabolite that is the ultimate product of this very gene. The seemingly simple inclusion of these modulators of gene expression introduces an enormous complication, as we discussed earlier, because instead of a linear chain of causes and effects, the Central Dogma has become a feedback system with different control loops. To make things even more interesting, rather than just one transcription factor per gene, there are often many, and the transcription factors themselves form hierarchical networks, where a high-level transcription factor controls the expression of numerous genes that in turn produce transcription factors controlling the expression of a whole set of other genes.

In addition to this feedback system, scientists more recently stumbled upon another fascinating control mechanism, which relies on hundreds, if not thousands, of small regulatory RNAs. Some of these have the ability to silence the expression of target genes, while others control how other RNAs are spliced together. As a consequence, it has become evident that small RNAs can be involved in a number of diseases. Thus, in addition to the feedback loops from genes to proteins and back, and from genes to metabolites and back, small RNAs form yet another loop from genes to RNAs and back.

Reductionism and the Central Dogma used to make biology look deceivingly simple, at least conceptually. Alas, if we really want to understand biology, health, and disease, the time has come to accept nature's complexity with all its beauty and challenges, while "making things as simple as possible, but not simpler," as Albert Einstein famously said. Life is complicated and it has become undisputedly evident that we need tools beyond our intuition to decipher its secrets. Such tools are at the heart of the emerging field of systems biology. Among them are experimental approaches that probe large and small biological systems and collect data and contextual information. Complementing these experimental approaches are mathematical and computational strategies. Some of these help us keep track of the many heterogeneous components of cells and organisms. Others have the goal of integrating biological information and of constructing models that permit exploration, explanation, and the formulation of novel hypotheses. Many of these models are initially conceptual, but as soon as they become more detailed and specific, they rely on the crisp language of mathematics, which alone is able to capture and evaluate complex numerical relationships among the components of biological systems. As we move into the future and learn more about living systems, the methods and approaches will certainly change, but the intricate features and characteristics of nature's complexity will always be with us.

2

I'd rather be fishin'

Biological research has had a long and esteemed history. So it is not surprising that its concepts, approaches, and methods have been subjected to dramatic changes time and again. Early trial and error in agriculture and animal domestication matured into simple plant manipulations and animal husbandry. Observations of birth and death, growth and decay, led to methods for preserving food for times of dearth. Exploratory dissections of corpses turned into primitive forms of surgery. The world-view of biology exploded with the invention of the microscope, which opened a window into an entirely new world of cells and microorganisms and pathogens. The exploration of medicinal herbs and poisons, as well as the procedures of alchemy and chemistry, motivated the invention of ever-more accurate methods and refined measurement tools.

The search for scientific truth reached a high point in the seventeenth century with the acceptance of the so-called scientific method, which is still considered fundamental today. According to this method, scientific inquiry advances through well-structured, iterative cycles of posing a hypothesis, testing it with experiments, analyzing results, making predictions, testing them, and formulating new hypotheses. In all fairness, one should mention that the roots of this structured type of scientific thinking and experimentation can actually be traced back two millennia to the third century BC Greek physician and anatomist Herophilus, who cofounded the most famous medical school of the time in the Egyptian city of Alexandria. Herophilus performed systematic dissections, which he documented in great detail, and maintained that trustworthy scientific knowledge can only be found on an empirical basis. Nevertheless, the scientific method became the gold standard only in the seventeenth century.

Then the twentieth century rolled along and modern biomedical research exploded. Powerful experimental tools and custom-tailored machines rendered it possible to characterize biological phenomena with a resolution never seen before, down to the level of individual molecules. A prominent highlight was the identification of the structure of DNA, but many other classes of molecule were identified and characterized, and uncounted small and large discoveries occurred during the second half of the century. Most of these breakthroughs resulted directly from the application of the scientific method, which brought forth incredible amounts of precise data and unprecedented insights into the inner workings of life.

In the shadows of this hugely successful modus operandi, an alternative approach began to take hold around the turn of the millennium. This approach was driven by novel combinations of molecular biology and ingenious engineering advances in miniaturization and robotics, which suddenly permitted the execution of very many experiments at once. Whereas it once had taken an entire thesis project to determine the sequence of a short gene, sequencing became a quick routine task. Quasi overnight, almost every molecular biology lab became enabled to characterize the expression levels of thousands of genes simultaneously, with no need for specifically targeting a gene of interest. It became feasible to identify hundreds of proteins or metabolites with techniques of mass spectrometry. Biology witnessed the birth of a new era of large-scale, high-throughput data generation.

While very exciting to many researchers, this type of investigation was seen by many others as the antithesis of the scientific method, a despicable distraction from real research. The idea of "let's see what happens if we check all genes" was derided as a "fishing expedition." But alas for the critics, fishing has been successful and often enjoyable throughout human history, and it quickly started receiving appreciation and acceptance within the biological science community. Fast-forward, and fishing for molecular targets is now largely considered an equal partner to traditional, hypothesis-driven research.

Of course, the new methods needed their own names, but unfortunately for everyone who likes words or is a linguist in disguise, they became collectively known as terms with the suffixes –ome or –omics. Thus, high-throughput data generation on proteins was named proteomics; for metabolites, it was metabolomics. Now, these suffixes seem to have become attachable to any and all biological quantities, leading to words like fluxome and pharmacomicrobiomics. Let truth be told, long before the unchecked proliferation of –omes and –omics, there were gnomes and comics, and now there are even gnome comics! And granted, the most prominent of the new breed of words, genome, is actually quite old. Not as old as the rhizome of plants, palindromes or economics, but at least older than Google and Wikipedia. The German botanist Hans Winkler coined the word genome in 1920, allegedly concocting it from "gene" and "chromosome." But the agedness of a word alone does not make the linguist shudder less, as "chromosome" is not composed of "chromos" and "ome" but of "chromo" and "some," from Greek "chrome" (color) and "soma" (body), referring to the heritable material in the nuclei of cells that could be seen as colored under a light microscope. However much they may cringe, though, linguists are not the ones advancing modern biology, and the flashy terminology is likely here to stay. So, –omes and –omics have generally assumed the meaning of an entire set of specific entities like lipids, giving rise to lipidomics. Still, one can only hope that some of these newly coined concoctions, such as omniomics (the all of all?), will eventually wither into oblivion and no longer be used by the humanome, formerly known as humankind.

Notwithstanding the terminology, or its character of a fishing expedition, –omics research has become fundamentally important in biology. In fact, one might go so far as considering it a brand new research strategy. It does not quite constitute a paradigm shift, because the traditional manner of reductionist biological thinking is alive and well, and will continue to be a major driver of biological advancement throughout the foreseeable future. Instead, reductionism and –omics research are now routinely being done side by side, and this dual approach will probably be the method of choice for the foreseeable future.

The novelty of –omics approaches becomes most visible in comparison to the reductionist process. As discussed in the previous chapter, reductionism reduces systems to subsystems and, ultimately, to their fundamental building blocks. Thus, it mandates precisely targeted research into components and their components at successively finer-grained levels. Absolutely nothing is wrong with this line of investigation, except for one implicit assumption, namely, that one can reconstruct a working system when all parts are available. This underlying assumption has turned out to be problematic, because some components do not function properly without their native contexts, and because groups of components may exhibit synergisms that cannot be attributed to any of the components by themselves. For instance, it is not possible to identify which single genes or proteins are responsible for oscillations that we observe in the cell cycle or in circadian rhythms. The impossibility is not due to insufficient experimental methods, but is genuinely driven by the interplay between several genes or proteins. In fact, if any component is removed from the system, it might just happen that the oscillations cease altogether. This emergence of unforeseen behaviors is of enormous consequence, because even the most rigorous investigation of the parts in isolation will never yield or even imply the interactive behavior. We have sequenced all the genes of some microorganisms, yet we cannot reliably predict the responses of these bugs to situations that we have not investigated before. We have a very good understanding of nerve cells, but we do not understand higher-level brain functions such as memory or cognition.

The philosophy of –omics research is very different. Instead of posing a hypothesis about some specific molecule or process, and then designing tests to support or refute the hypothesis, –omics research uses comprehensive measurements of the expression of many or all genes, the amounts of many or all proteins, or the concentrations of many or all metabolites, knowing full well that the vast majority of measurements will not be informative. This apparent "waste" is considered acceptable, because (a) the measurements are cheap and (b) it is quite likely that one will discover changes in genes or proteins or metabolites that nobody has ever considered before. In fact, we have seen over the years that many –omics studies pointed to genes or proteins whose roles and functions were entirely unknown. By its very nature, hypothesis-driven research would have missed these apparently important molecules altogether.

The number of simultaneous measurements in the field of –omics is astounding. The expression of tens of thousands of genes can be assessed with a single microarray experiment, and similar numbers of protein and metabolite species can be characterized at once with modern methods of mass spectrometry. As a consequence, our biological "parts catalog" is growing at a very rapid pace. It is no longer a total fantasy to imagine an essentially complete molecular inventory of a cell.

The ability to measure comprehensively has had an enormous impact on the way we approach biological problems. Indeed, it has changed the way we think about biological questions. Instead of formulating, sharpening, and refining a hypothesis before we start an experiment, we have the option of casting a very wide net in the hope that the results will guide us toward positing new hypotheses.

Driven by the technical nature of the experimental studies, a typical genomics study is exploratory and comparative. It may compare the expression of most or all genes of a normal ovarian cell with those of an ovarian cancer cell. The interpretation of such an analysis and its results is in principle simple: if there are significant differences in the expression of particular genes, these genes should be expected to have something to do with ovarian cancer. What exactly their role might be, we cannot derive from this experiment. It could be that they are causative or just symptomatic. It could also be that they are expressed as the cell's attempt to compensate for proteomic or metabolic changes triggered by the cancer. We just don't know. We do, however, have a new set of research targets. A follow-up investigation could be an experiment to silence or knock out one of the new candidate genes and to study the consequences. The experiment itself could be another –omics study or a more traditional, targeted study. This type of strategy has been employed thousands of times in recent years and identified uncounted numbers of genes whose expression is associated with a disease or other trait. In fact, the high-throughput nature of such genome studies has ushered in an entirely new field of genome-wide association studies, or GWASs. These molecular epidemiological studies examine common genetic variants within a population and try to deduce whether particular variants are associated with a phenotype, trait or disease.

In addition to identifying specific genes involved in a disease or trait, it has become possible to predict, at least to some degree, how a patient will respond to a risky drug treatment. A beautiful example in this new field of pharmacogenomics is a prognostic analysis for the treatment of acute lymphoblastic leukemia (ALL). Some children with this devastating disease respond very well to particular drugs, whereas others may exhibit such dramatic side effects that the population-wide efficacy and applicability of the drugs is severely compromised. Then, scientists at St. Jude Children's Research Hospital discovered with a genome-wide expression test that about 50 genes are expressed differently in children who respond positively or negatively to the drugs, and over 100 genes were identified in children responsive to one drug, but not others. It is to be expected that we will soon have large databases of such effects for most of the potent drugs, as well as their combinations, so that personalized genome diagnostics will be able to predict success, failure, or the nature and degree of side effects, before the drug is ever taken by the patient.

Corresponding high-throughput methods for proteins are not quite as advanced, but they are beginning to catch up. The situation with proteins is more complicated than with genes, because it is actually difficult to tally how many and which proteins are normally present in a particular cell and what their concentrations are. Furthermore, relatively minor alterations of the same protein, in the form of attached phosphate or sugar groups, can dramatically change the activity of a protein, and thereby its function. To make things even more challenging, the attachment and removal of such groups is very fast and often occurs as an immediate response to external or internal signals received by the cell. In recognition of these alterations, increasingly sophisticated techniques are being developed to characterize the activity states of proteins, and proteomics is in the process of becoming a very exciting and powerful approach to addressing changes in the cellular machinery under different conditions.

Similarly, metabolites can be measured quite comprehensively with methods of mass spectrometry, which can characterize tens of thousands of chemical entities in a small amount of sample. It is often the metabolites that are directly associated with a disease like diabetes, gout or even cancer, rather than genes or proteins, so that measurements of metabolites might provide better snapshots of health and disease. Then again, metabolite levels change dramatically during the daily cycle and are directly influenced by food, lifestyle, and many sorts of exposure.

All types of –omics data can offer very detailed insights into the status of a cell or organism. These insights are more telling if they are obtained as comparisons between normal and other samples or, even better, if they are generated at many sequential time points following some event, thereby offering an impression of the immediate responses of a living system to stresses or other stimuli. Especially informative are simultaneous assessments of the genome, proteome, and metabolome of the same system, because complex cellular responses often involve all three, and the results at one level can be connected to the results at other levels.

For the computational systems biologist, –omics data have enormous potential but also pose previously unknown challenges. Quite obviously, the results of these studies consist of very many data points that must be stored and handled. In fact, modern data files have become so complicated that they are often accompanied by detailed metadata files that describe how exactly the data were obtained. The field of bioinformatics has been addressing this complicated issue admirably over the past decade. A related issue with large –omics datasets is that many methods of statistics are no longer directly applicable, because –omics data cover many molecular species, but with relatively few replicates, which is opposite to the situation addressed by standard statistics. Thus, many new statistical methods for normalization and analysis have been under development since –omes entered the stage. A prominent example of such a method, called cluster analysis, is used to characterize objectively which groups of data are similar in their responses to some stimulus. For instance, if cells are exposed to some external stress, one might find that one subset of genes first increases in expression and then returns to normal, whereas another subset initially decreases and then recovers. A reasonable hypothesis then is that the different subsets of genes share a common role or function.

The statistical issues actually go much deeper, and tasks that at first appear to be straightforward are not so simple upon closer inspection. For instance, if there is a difference in gene expression between a normal and a diseased cell, how can we tell if the difference is really significant and biologically important? If a gene in a cancer cell is expressed twice as strongly as in the normal cell, is that significant? How about a difference of 10 percent? We clearly need a threshold or cut-off, but it is easy to imagine that the magnitude of the difference and its significance depend on the particular gene. A large difference may be needed if the corresponding protein is responsible for transporting materials in or out of the cell. Then again, flipping a control switch may not need much of a change in gene expression at all. The question of thresholds is confounded by various uncertainties and the innate variability in gene expression. Uncertainty derives from the fact that data are the result of an actual, physical experiment that usually requires numerous steps, each of which may introduce noise. Variability is due to the normal differences between two organisms, and it is becoming increasingly evident that even two cells of exactly the same type, within the same culture or tissue, may show very distinct characteristics. Taken together, the expression of a gene may be quite different between normal and cancer cells, but if the normal range within the human population includes both values, who is to say that the gene is important in cancer?

In addition to these vagaries, one must ask what the results of high-throughput genomic experiments really mean. Suppose a gene is highly expressed in a cancer cell but not in a healthy cell. If the role and function of the gene are known, one might be able to speculate why the expression is changed, but one cannot decide whether the change is causal or symptomatic. Even more difficult is the situation if the role and function of the gene are not known. While this situation directly creates a novel hypothesis, the result by itself is difficult to interpret. It is moreover not always clear that an expression change in a gene actually corresponds to a similar change in the protein for which it codes. As a consequence, changes in gene expression are important, but somewhat indirect markers. Finally, very many items are measured in a typical –omics experiment. It is possible that some of the measured items show altered expression or amounts, which merely compensate each other under stress conditions or disease, and that they are not causative at all. Also, most data represent averages over thousands of cells, which may mask important responses occurring only within a few cells.

Fishing has had a fast and illustrious career start in the brave new world of –omics. Not even in existence a couple of decades ago, it endured much ridicule during its infancy, only to emerge as one of the most important launch pads for investigations of large and complex biological systems. Ideally, such investigations proceed in the following manner. Several –omics experiments are performed at the gene, protein, metabolite, and physiological levels, in a comparative manner, and possibly even in time series. By themselves, the resulting data are so overwhelming in quantity that they cannot be interpreted. However, as will be discussed in the following chapters, methods of bioinformatics and computer science organize these data and uncover patterns within the ocean of information and noise that are invisible to the unaided eye, but nevertheless reveal critical associations, for instance, between gene expression and disease. These associations, in turn, lead to novel hypotheses that may be tested experimentally or are entered into computational models whose goal it is to integrate the information in such a manner that it offers explanations and leads to true knowledge.

3

Whizzes and apparitions

The rapidly emerging high-throughput data generation of the new world of –omics has radically changed the way we approach many questions in biology. The traditional approach had called for a painstaking, step-by-step process of formulating an hypothesis, testing it with well-designed and well-controlled experiments, analyzing the results for new insights, interpreting these, and posing the next round of hypotheses. In stark contrast, a single –omics experiment generates thousands of data points, and the researcher knows full well that most of them will not advance our current state of knowledge. Yet, there is justifiable hope in finding the proverbial diamond in the rough or maybe at least the needle in the haystack.

Because it has become relatively easy to perform –omics experiments, the amount of biomedical data generated every year has become overwhelmingly huge, leading to the new buzzword of "BigData" which vaguely refers to datasets that are so large and complicated that traditional tools of database management, analysis, and interpretation are no longer workable, at least not with any degree of efficiency. In the case of –omics, the big data experiments often come from interdependent collections of different types of experiment – such as genomics and proteomics, performed on the same system, maybe on different days, and maybe under different conditions or stimuli. The rationale for these combined approaches is the expectation that parallel datasets, elucidating different aspects of the same phenomenon or disease, should be much more informative than datasets of the same type, because they complement each other. For instance, one would expect to see some changes in gene expression reflected in the amounts of various proteins. As a consequence, the combined datasets are huge in size and heterogeneous in nature. For the traditionally trained biologist, this means that "double, double toil and trouble" are brewing in the world of modern biology, leading to ever-louder SOS cries begging computer nerds for help. In many of these situations, the computer whizzes are asked for apparitions: please make patterns appear in the vast expanse of our data!

Computer whizzes actually like this type of challenge, because it allows them to play with bigger and faster machines that can handle numbers of bits and bytes whose prefixes only hard-core scientists have even heard of, such as peta–, exa–, and yotta–. As an extreme example, imagine recording all the data that are currently measured on the Large Hadron Collider, the super-high-energy particle accelerator in Geneva of Higgs boson fame. The daily amount of data would be about 500 exabytes, which equals 500 quintillion, a five with twenty zeros, that is, half a billion trillion. Biology does not produce as many data by far, but finding information in the richly flowing data stream sometimes does seem like the proverbial drinking from a fire hose. We could step away from the fire hose a bit and gain some distance, but then we might miss the diamond we covet.

While computer whizzes like the challenge, they cannot just fire up the cauldron and cool it with baboon blood, as Shakespeare's witches suggested in *Macbeth*. Instead, they have been working incessantly on computer algorithms of many new varieties, with some specifically designed for dealing with BigData. Many of these algorithms are mixtures of statistics and computer science and fall into the general categories of artificial intelligence, machine learning, pattern recognition, and data mining. Their overarching goal is to scan large datasets for associations and to reveal patterns that are not recognizable directly by the unaided human mind.

One of the pioneers of machine learning, Arthur Samuel, defined this methodology in 1959 as "giving computers the ability to learn without being explicitly programmed."[1] This definition deserves closer attention.

1 Cited in P. Simon (2013) *Too BIG to IGNORE*, John Wiley & Sons, Hoboken, NJ.

Typically, computers are programmed to solve some numerical problem, such as computing the average of 10,000 numbers. The task is very specific, and the programmer writes computer code that corresponds to the way a mathematician would attack the problem: add all the numbers together and then divide by 10,000. In machine learning, the tasks are much more ambiguous. A typical question might be: do the data contain measurements that help us distinguish healthy cells from cancer cells? It is not that we ask the computer to sort the data and spit out a subset with known features. No, we ask the computer to search for those data that, possibly in complicated combinations, have something specifically to do with cancer, whatever that might be. For tasks of this nature, the programmer typically does not even know a good strategy, let alone a formula. So, how can a computer do it?

If computer algorithms find good solutions, they appear intelligent to us; just think of IBM's Deep Blue computer that beat several chess grandmasters. Intelligence is difficult to define, but it is generally accepted that machine learning is a subset of the larger field of artificial intelligence. Machine learning was ushered in by the famous British mathematician and computer scientist Alan Turing in the 1940s and 1950s. He famously pondered and mathematically analyzed the question, *Can machines think?* This is not only a complicated but also very loaded question, and because words like "machine" and "think" can be interpreted in many ways, Turing suggested a game, later called the Turing test, in which a human judge asks questions that are answered independently by another person and by a computer. If the computer could become so sophisticated that the judge could not tell which answer came from the person and which from the computer, one would have to admit that the computer had the ability to think. Thus was born the field of artificial intelligence and machine learning.

Since Turing's times, artificial intelligence has become an enormously important field of research, and applications are found not merely in science but also throughout daily life, from our encounters with Google or Walmart to the sleuthing activities of governments worldwide. One of the early methods was pattern recognition, which prominently became able to classify handwritten addresses on envelopes according to zip codes, thereby making mail-sorting thousands of times more efficient than in the olden days. In modern times, similar methods filter out spam email and tell car dealers that we had been browsing customer surveys of cars. Pattern recognition methods provide answers to classification problems, and if a handwritten 4 looks like a 9, these methods can even tell us what the likelihood is that their inferred answer is correct.

The key ingredient for computers to learn is actually not all that different from the human experience. It consists of "training" in the form of a dataset from which the computer learns what to do. For instance, it may learn from large datasets on traffic accidents that seatbelts and airbags save lives. Once the machine has learned an association, it may be enabled to generalize this learning experience toward new scenarios. Like humans, the more experience the machine amasses, the better it becomes equipped to assess similar, but new, situations.

Two classes of learning method are of particular note in systems biology. They are called supervised and unsupervised learning, respectively, and both are important. In supervised learning, the computer is first provided with training data. For instance, one could provide the computer with the results from a large epidemiological study that measured the typical culprits of weight, age, gender, blood pressure, cholesterol levels, smoking habits, and similar features, along with many other data. Of special note for supervised learning is that the training data also contain the information that individuals in the study had already been diagnosed with some disease; let's call it D. All data are entered into the computer, and the machine learning algorithm tries to identify associations between disease D and some of the other types of measurement. At the end, the algorithm might "predict" that the prevalence of disease D is substantially higher in one specific subset of people than in the rest of the population. For example, these high-risk folk could be smokers of low weight and high blood pressure, who are also carriers of some gene that we might call G1234. In addition to this identification, the algorithm gives each of these criteria a weight that corresponds to its importance for the disease. In the second "validation" phase, the same algorithm is given data on another cohort of people, who had already been tested, but whose data were not used for training. Then the algorithm is asked to make a prognosis, based on the same criteria of low weight, high blood pressure, gene G1234, and smoking, of who is at high risk of developing disease D in the near future. The quality of the learning algorithm is assessed based on its generalization error, that is, on the percentage of the new data that the algorithm classifies correctly or incorrectly. Internally, the successful algorithm has in the process established an association model that can classify new populations into those individuals who are likely or not so likely to develop disease D. While the algorithm magically categorizes the population, the results do not provide any explanation for why gene G1234 might predispose a person to disease D.

The most widely known learning algorithm is the artificial neural network, or ANN as it is lovingly called in the field. It dates back to the 1940s and was inspired by the way the brain works, although the similarities are quite superficial. The principles of the analogy are not difficult to understand. ANN consists of layers of "neurons," which are really mathematical functions, and "synapses" that functionally connect the neurons. An input layer of neurons receives a stimulus. Each neuron in this layer is connected through synapses with all the neurons of a second layer, and these are in turn connected to all the neurons in an output layer. Sometimes there are more than three layers, but three is the typical minimum. Each connection between neurons is assigned an importance score, which consists of a numerical value or "weight." Initially these weights are all the same or they are randomized. The basic learning mechanism of this layered array is the following. Any input to ANN "fires" up some of the very many neurons in the input layer. For instance, the neurons in the input layer could respond to light-and-dark sensors that "see" a postal code on an envelope, even though they have no clue what the pattern is or means. The excited neurons fire up neurons in the second layer. They do so with intensities that correspond to the current numerical weights of the connections. The same happens between the second and the output layers. The degree of firing among neurons in the output layer is evaluated and interpreted. If ANN arrives at the correct answer, for instance by revealing the correct postal code, it is "rewarded." Specifically, all connections between neurons in the different layers that were involved in this round are altered such that they become more influential the next time around. Connections associated with wrong answers are penalized and become less influential in the future. The output of the neural network consists of some classification. In the simplest case, it is binary, for instance referring to health or disease, or it could be the correctly or wrongly identified number on the envelope. However, the output could also be much more complicated. In fact, the sky is theoretically the limit.

During the training phase, the algorithm is rewarded for correct answers and penalized for incorrect ones. This system is implemented such that each correct input–output relationship strengthens the connections between the neurons in the three layers by increasing their numerical weights. Wrong responses are penalized by weakening the connections. This reward–penalization system is called back-propagation. Its details permit numerous variations that make specific implementations of ANNs particularly efficient.

Intriguingly, sufficiently large ANNs have the capacity to learn any input–output relationships, no matter how complicated they are, if the algorithm trains long enough. The greatest appeal of ANNs stems from the fact that the computer learns complicated relationships all by itself and exclusively from the training data. In particular, it does not require the researcher to infuse prior biological knowledge or specify any application-specific mathematical functions.

As is to be expected, there is seldom anything for free. Here, a substantial drawback of this and other machine learning methods is that a reliable generalization or prediction for new data may require very large, representative datasets for training, which are not always available. The more complicated the true input–output relationship is, the more training data are needed and the larger the neuron arrays have to be.

In biology and medicine, ANNs have been used in vastly different applications. Many applications can be found in the field of cancer diagnostics, where quite reliable disease predictions and patient outcome prognoses were achieved by combining different clinical measurements or analyzing retrospective data from patients who had received a particular therapy. Other applications have classified medical microscopy images with the goal of cell type classification. ANNs have also been used in the –omics sciences, for instance to classify gene expression patterns and protein and metabolite profiles.

A different type of algorithm for supervised learning, which has become popular in recent years, is the support vector machine. The name is somewhat misleading, because the machine is really an algorithm. The algorithm presents the data as points in a high-dimensional space and attempts to classify them, for instance into diseased and healthy individuals. This task is accomplished by optimizing a mathematical function that cuts the space such that (almost) all points associated with disease are on one side and (almost) all points associated with health are on the other. The optimized separation function can subsequently be used to make prognoses on new data. Support vector machines have been applied to large genomics, proteomics, and metabolomics datasets, to protein interaction networks, as well as to the prediction of the structure and function of proteins and the identification of target sites for the translation of RNA into proteins. For the analysis of gene expression studies, support vector machines now rank among the most efficient algorithms for identifying gene clusters that are similarly expressed in response to different stimuli.

The yang corresponding to the yin of supervised learning is unsupervised learning. Here, the goal is not to make generalizations based on input–output relationships, but to discover unknown types of hidden patterns in the data. The most important difference is found in the training data themselves. In the previous example, the training data contained information regarding which individuals had already been diagnosed with disease D, and this information was used to train the algorithm. In unsupervised learning, information on disease D is not available, and it is not possible to provide or make use of feedback. Instead, the unsupervised algorithm tries to identify patterns in the data that characterize aggregates of associations. A typical and important application in biology is cluster analysis, which attempts to sort data into groups such that data within each group have more similarity than data from different groups. The meaning of this similarity depends on the application. As a prominent example for –omics studies, the algorithm could be tasked to group genes that exhibit similar expression changes in response to an array of different stimuli. The rationale for this task is that the genes in each group might be related in some sense. For instance, they might be regulated by the same transcription factor or code for sequential enzymes in the same pathway. Another application is the phylogenetic characterization of the evolutionary similarities and relationships among different species. Many clustering algorithms have been devised over the past two decades.

All methods described so far contribute to a new field that has been given the buzz-term data mining. Data mining has as its goal the extraction of information from large datasets and the characterization of structure or patterns within the data. The hope is that this structure can later be used for other purposes, such as predictions or disease prognoses. For purists, the terminologies and distinctions are not quite crisp enough, but one might say that machine learning makes inferences from already known properties of training data, whereas data mining aims to discover formerly unknown properties of the data. Data mining is therefore an integral component of the tasks of knowledge extraction and knowledge discovery from datasets. All these efforts aim to create abstractions of the input data, in that they distinguish key information from tangential information and noise. Conceptually, data mining is not all that different from a regression analysis in statistics, where many data points are reduced to a regression line that can be used for predictions of future events or correlations. In contrast to this more traditional data analysis, data mining deals with very large datasets with many dimensions, is automatic or at least semi-automatic, and attempts to discover something entirely new that so far has been hidden from our perception. In addition to cluster analysis, data mining may detect anomalies as well as complex dependencies among data. It uses information from data of the past to forecast responses to stimuli that might occur in the future. One important aspect of the abstraction gained in data mining is a compact representation of the data, which secondarily permits a simpler visualization of the structure that was formerly hidden in the data.

Two important, genuine features of essentially all machine learning algorithms are at the same time their greatest assets and their most significant limitations. First, all machine learning algorithms are based purely on associations. The advantage of this feature is that not much input is needed outside of training data. The drawback is that these algorithms do not provide any rational explanations, even if they are very good at predicting the outcome of a new experiment. The "data model" they provide is a set of numbers, such as the weights in an optimized artificial neural network or the numerical features of the separation function in a support vector machine. Beyond the specific setting for which these numbers were computed, they have no real meaning or interpretation. Second, machine learning methods may be arbitrarily complicated. For instance, an artificial neural network may have an essentially unlimited number of neurons. The more neurons it has, the better it will be able to represent complicated input–output relationships. Thus, these algorithms are very general and potentially very powerful. At the same time, if too many neurons are used, the patterns in the training data may be perfectly matched, but analyses of new data utterly fail. This so-called overfitting problem is due to the fact that the training data were not sufficiently informative for the analysis that was attempted, or that the algorithm tried to fit features that were due to noise in the data.

Even if machine learning, data mining and data reduction methods have their issues, they are often reliable and predictive and therefore serve as the first line of defense in the analysis of BigData. Furthermore, methods like cluster analysis tease possibly unknown associations out of the data, and although these associations do not imply causality, they can provide much-needed guidance for the formulation of novel hypotheses regarding possible chains of mechanisms that lead from a stimulus to the observed response. Because they can hardly be obtained with any other means, these hypotheses are often very valuable. In many cases, they may be tested directly with traditional experiments, thereby propelling the field forward. Indeed, this is the modus operandi for many biological labs of the twenty-first century. The new hypotheses may also be reformulated as targeted questions that can be explored with a mechanistic computational model, which has the potential of leading to its own new insights.

Taken together, the apparitions extracted by computer whizzes from −omics data may not yield explanations and insights by themselves, but they often become well-justified launch pads for further experimentation and for the construction of mathematical models.

4

Why?

"Mommy, why is the sky blue? Why do zebras have stripes? Why has Aunt Maud got hair growing out of her nose? Why are bananas crooked?" All good parents try hard to answer such questions, or at least most of them, but it is hard-core scientists who most encourage their kids to ask a lot of questions. After all, questions are a sign of curiosity, and curiosity is the ultimate, indispensable prerequisite for becoming a good scientist. Yet, as soon as the same scientists return to their labs, "why" becomes a taboo word. Why? What happens on the way to work?

Many scientists subscribe to the tenet that "why?" is not a scientific question. "How?" and "what?" are scientific questions, but "why?" is not. The famous evolutionary biologist Richard Dawkins even derided it as a "silly" question. One reason for the discrepancy is probably that children are usually satisfied with a single answer. "Bananas grow faster on one side than on the other" might just do. No need for a lecture on phototropism or auxins or other phytohormones that play a role in coordinating growth processes in plants. Not so with scientists. Every answer leads to new questions, and while many in the chain may have answers, there comes the inevitable point at which one has to admit "I don't know" or "nobody knows," or a supernatural deity is to be evoked quasi as a *deus ex machina*. None of these three options is particularly desirable to scientists, and "why" is therefore *a priori* to be excluded from scientific discussions as a preemptive measure. Answering "why" implies that biological processes are driven by intent toward a goal; it reeks of teleology, the many-centuries-old philosophical concept that searches for the ultimate reasons of being and the purpose of human existence. It also comes uncomfortably close to questions of creationism and intelligent design, which are counter-scientific because they do not admit testable hypotheses.

If we take a detour around the semantics of the "silly why," it does not take long to discover that scientists really do look for explanations. Thousands of studies have searched for causes of cancer or other diseases, and the investigation of just about every mechanism in biology tries to answer a question of causality. In fact, chains of causes and effects are the most prominent means of explaining biology. They help us understand what we observe and guide our strategies of developing and formulating new scientific hypotheses. Therefore, while avoiding "why," we should ask how these explanatory chains are discovered. The answer, of course, depends on numerous factors, ingenuity, serendipity, and the state of knowledge. Nonetheless, one may distinguish two paradigms that are dominant in systems biology, and that ideally complement each other.

The first line of searching for explanations begins with large –omics datasets, which were discussed in the previous chapters. The generic strategy is the following. Cast a very wide net and measure as much of the system as feasible. For instance, investigate a phenomenon by measuring genome-wide gene expression, the abundances of all major proteins, and the concentrations of most metabolites, either as comprehensive snap-shots or as several series of measurements over time. These experiments generate thousands of data points, a fact that of course is desirable, but also creates a problem, as lots of irrelevant information and noise may cause us to miss the forest for the trees. The conundrum suggests employing computational methods of machine learning and artificial intelligence, as we already discussed. These methods have the capacity to detect association patterns that are possibly very complicated. They may be invisible to the human eye, but they don't evade rigorous com-putational analysis. Such patterns might connect gene expression with changes in specific proteins and metabolites or may yield novel com-binations of over- or under-expressed genes and proteins that, often for unknown reasons, seem to be associated with a disease or some other system output of interest. Importantly, these associations may well be statistically significant, but they do not necessarily imply causality. One may be the cause of the other, but both could also be the consequences or symptoms of some other underlying causes. For instance, a disease may be caused by a single gene mutation, as is the case in cystic fibrosis and sickle cell disease, but it also happens quite frequently that a disease causes many genes to be differently expressed as compensation for other disease-related problems. In fact, many genes of cystic fibrosis sufferers show altered expression, but only one gene is the true cause of the disease.

Distinguishing association or correlation from causation is important, but by no means easy. In fact, the search for causes and effects is probably as old as mankind and has intrigued thinkers throughout the millennia. Aristotle grappled with the issue, stating that a cause must precede an effect. Galileo went one step further. In his 1638 book, *Two New Sciences*, he agreed with Aristotle's requirement, but postulated in addition that a true cause must be both necessary and sufficient. The nineteenth-century English philosopher John Stuart Mill added yet another requirement, namely, that one had to eliminate other possible explanations of a cause–effect relationship. In the early twentieth century, the American geneticist Sewall Wright formulated causes and effects in the language of linear algebra, and his concepts became the foundation of modern causality analysis. Finally, Clive Granger, the Welsh 2003 Nobel Laureate in Economics, proposed a statistical metric stating that, if A is supposed to be causing B, then measurements of A should contain significant information regarding future trends in B.

In theory, the mathematical and statistical metrics are valuable, but, in practice, causality is often very difficult to prove, for various reasons. First and foremost, the causes may be indirect, and there could be alternative contributors to an observed effect. Moreover, most systems in the real world contain numerous closed cause-and-effect loops, à la chicken and egg, and this type of circularity makes statistical methods problematic and the search for true causality troublesome. Ultimately it appears that a practically and generally applicable theory of causality is still missing, which is quite amazing, as we have been searching for over 2,000 years.

A common modern approach toward causality pursues a much simpler task. Namely, one asks whether two variables are mutually dependent. A typical example in genomics is the following. Consider two genes that appear to be similarly expressed in many situations. That is, if different stimuli are applied to the cell or organism, the expression of both genes usually changes in the same direction. The question is then whether we can make reliable predictions about the expression of one of the genes in so far untested situations, if we know the expression of the other. If so, we can further investigate whether the two co-expressed genes are functionally related, for instance, by coding for enzymes within the same pathway. This type of "mutual information" has the great advantage that it can be assessed with rigorous methods of probability theory and therefore provides an unbiased tool for quantifying the relatedness and interdependency of two variables.

To summarize, the first approach to establishing causes and effects in systems biology works from the top down in an initially exploratory fashion. It starts with large datasets, which in their raw form are incomprehensible for the human mind. Machine learning establishes patterns among some of the data, which might imply associations between combinations of certain measurements and an output marker. To some degree, one may be able to establish causality, but more often than not this is difficult if not impossible. Are we stuck, or can we convert association patterns into explanations?

To the rescue often comes the second, complementary, approach of searching for explanations. It is anchored in the traditional biological thinking of discovering and characterizing parts and processes of systems. The goal of this approach is to uncover the mechanisms with which nature solves a specific task, and ideally, these mechanisms can be explained in terms of the first principles of physics. Even if such a reduction is not completely possible, one can still be satisfied with explanations that characterize the role of each system component for the functionality of the whole. As an example, we might not be able to reduce the function of a ribosome entirely to fundamental physical forces. Nonetheless, science has clearly and in detail established the ribosome's capacity for translating RNAs into proteins at the molecular level.

In a coarse simplification, this mechanistic approach is sometimes called bottom-up, because it attempts to reconstruct systems – and models of systems – from their components. Indeed, the vast majority of modeling successes so far have resulted from this type of modeling, which represents all processes individually and then connects them in the way the biological system is assumed to be connected. However, even bottom-up modeling usually does not start at the very bottom, but rather at some reasonable "mesoscopic" level within the hierarchical biological organization. The key difference to the exploratory top-down strategy is that this approach is hypothesis driven. Typical questions are the following. Are the observed or hypothesized processes sufficient to connect the phenomenon of interest functionally to other parts of the system? Can we infer what is missing from our current understanding? Do the data and observations allow us to identify and fill gaps in the chain of causes and effects? Do the known parts and processes allow us to formulate novel hypotheses that, if confirmed, will help us refine or even complete the picture?

The characterization of parts of a system and the functional reconstruction of this system from its parts are accomplished with targeted experiments and logical reasoning. This reasoning integrates what is known, identifies unresolved issues, and formulates crisp hypotheses that address specific aspects of or gaps in the current understanding of the system. Experiments confirm or refute these hypotheses and by doing so help fill the gaps or lead to further hypotheses. This strategy of the scientific method has been extremely successful. However, the approach is naturally limited to a focus on relatively small or simple systems, because the logic behind the experiments relies entirely on the human mind, and our mind has quite a limited capacity when it needs to merge and integrate complicated numerical relationships. Maybe computers with effective reasoning algorithms will guide systems biology in the future but, at present, the logic behind experiments depends on the ingenuity of researchers.

Other chapters discuss the challenges posed by complex systems to the unaided human brain in more detail, but a few examples are sufficient here to revisit the issue. An almost trivial scenario that nevertheless is difficult to resolve is a process that is simultaneously inhibited by some factor and activated by another. If we know that both factors are present, how can we foresee whether the process will be active and, if so, with what strength or magnitude? Other very prevalent scenarios contain feedback signals, networks of regulation, or thresholds whose effects depend on the numerical characteristics of the system, which our brains simply cannot assess appropriately.

Thus, the traditional reconstruction of biological systems, using experimental means and mental logic, has been a terrific tool for assessing "local causality," but is not necessarily sufficient to address systemic causality. Particularly challenging systemic causalities are associated with "emergent features," which cannot be uniquely related to any one of the components of the system. As a poignant illustration, consider a specific differential equation that can generate chaotic output. It is aptly called a blue-sky catastrophe because, out of the blue and without intervention or warning, the initial oscillations generated by the equation become erratic and unpredictable. The differential equation contains a parameter value, which must be set in just the right numerical range to lead to this type of chaos. Outside this range, chaos never occurs, and the differential equation generates a nice periodic oscillation. Does the parameter "cause" chaos? Clearly, it contributes to it, but the parameter alone can hardly be responsible. Instead, it is the whole of the system that may or may not lead to chaos. Because we cannot really explain such a behavior, the only currently available tool for objectively and quantitatively assessing systemic causality and the emergence of new system responses is computer simulation. This important topic is the subject of Chapter 9.

The question of "why?" has recently made some sort of comeback in a specific domain of systems biology. Researchers within this domain are searching for design and operating principles that nature uses time and again to solve specific tasks. One of the pioneers of modern systems biology, Michael Savageau, has been spearheading this line of research, arguing the following. If there are no general rules or principles that nature uses, then every system could be genuinely different, and every systems analysis must start from square one. By contrast, if general rules and principles exist, and if we can identify them, then the analysis of a system we had never encountered before is greatly constrained, and we can immediately build upon the general concepts and test default ideas regarding how this system might function. An analogy is an experienced watchmaker, who is often able to repair a watch even if he has never seen the particular brand before.

To intuit the power of knowing rules or principles, consider the genetic code. It says that triplets of particular nucleotides in DNA or RNA code for specific amino acids. The code is very close to universal and covers viruses, bacteria, plants, and animals of all stripes, including humans. The enormous advantage of knowing the code is that we never have to establish it again. When we investigate a new bacterium, we determine its gene sequence and know immediately which amino acids are represented in its coding regions. Using this information, we can make strong inferences regarding the proteins in this new bacterium and, thereby, also its functionality. If there were no common coding scheme, we would have to determine the specific code for every species anew.

The search for design principles is motivated by this power of generality. As a beautiful example, Savageau covered an entire class of apparently different biological phenomena by deciphering rules that govern the regulation of gene expression in microbes. It had been assumed that the mode of regulation was happenstance, but Savageau demonstrated that, under certain, well-characterized environmental conditions, only activators are used, while under different conditions, repressors are employed. Similarly, his group analyzed patterns of feedback regulation. Asking the dreaded "why?" question, they searched for the rationale of the common observation that the last product in a metabolic reaction chain inhibits the first step. Many alternative patterns of regulation could be imagined, but this one is chosen by nature time and again.

Over the past four decades, Savageau's group and others have been developing rigorous mathematical and statistical methods to analyze and answer the question of why nature apparently solves particular tasks with one preferred design rather than with alternative designs. In fact, such questions regarding the design and operation of biological systems have become quite popular in recent years. They are addressed with a two-phase investigation. In the first phase, one considers a specific phenomenon, such as the structure of a regulatory circuit, and tries objectively to identify the rationale for this structure. The main question in the second phase becomes whether the same structure or pattern is being used over and over again in a narrow niche or in a wide array of biological systems.

"Why" is a very useful interrogative, even in science, so long as it is not misused or misinterpreted. Philosophers and laypeople alike have been struggling with the meaning and purpose of life, and even deep thinkers have not found ultimate causality. If they ask "why?" in this context, science cannot offer much help. It does not answer the questions of why we are here or what came before the Big Bang. Nevertheless, the principal activities of science are discoveries about and explanations of the world surrounding us, and whether we ask "what is the rationale for" or "why" appears to be more of a semantic issue than a fundamental problem.

5

Simply engenious!

Whether it is acting, race car driving, or exploring the moon, it is usually the actors, drivers or astronauts who make the covers of magazines and receive celebrity accolades. And most of them certainly deserve praise for their hard work and the perfect execution of the task at hand. But we all know that their rise to stardom happened on the shoulders of uncounted others whose names flash by in very small font at the end of a movie or are not even mentioned. If we consider car racing, for example, a huge number of individuals contributed to the creation of the infrastructure for the complex system within which an Indy 500 victory is possible. Engineers not only designed and built the cars, but also thought up and actually realized the idea of tough yet light helmets, fire-resistant suits, pits custom-made for speed, the race track itself, and the entire support system allowing drivers to race, spectators to cheer, and venders to sell their wares. And so it happens in many situations that the thinkers and tinkerers and organizers, the makers and builders behind the scenes, remain in the shadows and out of the public eye, even though it is they who make miracles possible. To some degree this is not surprising, as engineers, mathematicians, computer scientists, and ingenious nerds of various types are not notorious for their interest in social hobnobbing or braggadocio. The situation in systems biology is not all that different. Here, it is experimental biologists or clinicians, founders of biotech start-up companies, or producers of novel medicines, who may receive at least a bit of attention from the public, and whose success rests firmly on the shoulders of uncounted, unsung heroes from the field of engineering.

The most dramatic example is arguably the paradigm of experimental systems biology itself, the –omics revolution, where widely noticed insights into the inner workings of genomes, micro-RNAs or protein interactions have only been possible due to ingenious engineering that led to miniaturization and permitted the high-throughput execution of many parallel experiments with robots. Unluckily for their image in the public eye, these robots are nothing like *Star Wars'* R2-D2; they don't smile, they don't look sad when scolded, and they do not shake hands. They are sterile glass boxes with motors, tubing and pipettes, whose looks on the covers of popular science journals would hardly inspire the imagination of curious youngsters. Yet, these robots have become indispensable tools for systems biology. They are very precise and efficient, loading minute biological samples into tiny wells on specialized plates, and sending them to their –omics analysis, sometimes 24 hours per day, without much direct supervision, and without ever asking for a raise or vacation. Imagine that! Just a few decades ago, the sequencing of a gene was the topic of an entire doctoral dissertation; now high-throughput machines can sequence essentially the entire human genome for about $1,000 within a timespan that is measured in days if not hours.

Engineering has been a crucial driver of progress in biology and medicine for a long while, and uncounted examples documenting some of the incredible advances range from artificial limbs and heart valves to novel biomaterials that provide scaffolds for cells as they regenerate lost tissue; from the earliest light microscopes to atomic force microscopes that allow imaging and manipulations at a scale that renders the exploration of mechanical properties of individual cells possible; from fuzzy X-rays to amazingly detailed CT scans; from specialized tools in the operating room to methods like laser capture microdissection, which permits the isolation of single, living cells from tissue; from applied control theory to sophisticated interfaces between nerve cells and computers, which are beginning to allow the operation of prostheses purely through thought; and from simple pacemakers to flexible electronics etched on plastics that are stiff as glass at room temperature but become very soft and malleable when inserted into a brain for stimulation or probing.

Specifically with respect to systems biology, it is evident that any system can only be truly analyzed if at least a good portion of its components is known, whether these are different species of an ecosystem, various cell types in blood, or distinct molecules within a cell that all have their own roles and functions. This characterization of a biological system inventory often occurs in overlapping phases that attempt to answer three questions: What is there? How much of it is there? Where exactly is it located? In medieval times, anatomists were able to find answers to questions about the human body after an illegal midnight trip to the cemetery. In our modern times of –omics, answers are less dangerous but also harder to come by. After all, a cell, which is typically a few micrometers in diameter, contains thousands of different types of molecule. Rising to the challenge, engineers have been coming to the rescue time and again. In close collaboration with biologists, they have been designing amazing techniques for molecular detection, quantification, and imaging, and have been building increasingly sophisticated gadgets to support these techniques.

Machines and methods for establishing and quantifying molecular inventories must be able to distinguish many different types of metabolite or protein that are all mixed up in cells, blood serum, cerebrospinal fluid, urine or some other biological sample. Often they have ominous-sounding names such as MALDI-TOF-MS, which become even more ominous when spelled out: matrix-assisted laser desorption/ionization time-of-flight mass spectrometry. Most certainly a hot conversational topic at any dinner party! Simply speaking, a MALDI-TOF-MS analysis requires a very small volume of a biological sample, which is positioned in the MS machine on a specifically designed artificial surface, called a matrix. A laser is shot at this matrix, causing molecules from the sample to be blown off (desorbed), electrically charged (ionized), and then propelled toward a receiver. The time of flight to reach the receiver is measured. This time depends on a molecule's mass and is therefore characteristic for each particular type of molecule. The output recorded by the machine consists of a spiky pattern (spectrogram) of the different masses. The most astounding feature of this technique is the stark contrast between very large and very small: a high-quality MALDI-TOF-MS can distinguish tens of thousands of different molecular species, even if they are very similar. At the same time, the smallest amount of a given molecule that can still be measured may be in such small units that our common language does not even know their names. For instance, MALDI-TOF-MS can measure pieces of proteins in the attoliter range. An attoliter is 0.000 000 000 000 000 001 of a liter; it takes roughly 50 trillion attoliter droplets to fill a raindrop!

MALDI-TOF-MS is only one example from among many. Microarrays allow us to measure the expression of tens of thousands of genes in a single experiment. Nuclear magnetic resonance (NMR) makes it possible to measure several metabolite concentrations within cells without killing or even harming them. Flow cytometry permits the sorting of cells with slightly different features at a rate of tens of thousands per second. These features may be very subtle; for instance, cells may be sorted based on slightly varying proteins on the outside of their membranes. Mass cytometry makes it possible to measure the activity state of dozens of signaling proteins inside single cells. These amazing techniques, along with numerous variants, allow the systems biologist to establish incredibly detailed molecular inventories and to quantify them, at least to some degree.

It is not only important to know what and how much is there, but also where exactly specific molecules are located. Again, one can only marvel at novel technologies, just as people in 1895 marveled at the new X-rays, for which Wilhelm Conrad Röntgen received the very first Nobel Prize in Physics in 1901. How could it be possible to "look" through flesh? Today, we can see not only bones but also subtle differences in tissues, including tumors, in an entirely non-invasive manner with CT, PET, and MRI scans. Variations of these methods even render it feasible to image features within a living brain; as an example, functional MRIs can identify the electrical activities of specific brain sections that are involved in the execution of a particular task.

Modern technologies allow us to explore much finer details, incredibly, down to individual protein molecules. These can be seen when they are tagged with a marker that lights up under UV light or some other defined wavelength. The most famous among such markers is GFP, a green fluorescent protein that was first encountered in jellyfish. Its gene can be tagged to the gene of an interesting protein of a host organism, and when the host expresses the DNA as a protein, GFP is expressed as well and lights up under UV light. The green glow indicates that the DNA of interest was transcribed and translated into a protein, and modern microscopy can show where this protein is located. Many such markers are available now in different colors, permitting the study and stunning visualization of the locations of proteins, molecular structures, and even viruses that have infected a cell. It is also becoming feasible to create specific artificial molecules, called quantum dots, inject them into the bloodstream of a mouse, and let them seek out tumors, which can then be visualized under UV light, in some cases without harming the mouse. A sophisticated variation of MALDI-TOF-MS makes it possible to scan a section of tissue and to establish a two- or three-dimensional record of different molecules at every point of that tissue section.

When it comes to biomedical systems, size matters greatly; small size that is. Cells are small, and bacteria and viruses are much smaller, so that any type of miniaturization is a very beneficial feature for their investigation. Indeed, the term du jour is *nano*, the Greek word for *dwarf*. In modern terminology, nano means *really small*; in the metric system, it refers to one-billionth. While the prefix nano can be found in many contexts, not just in the sciences, its definition actually becomes fuzzy for biomedical systems, because most molecular events occur on a nanoscale anyway, so that there is little difference between molecular scales and nanoscales, except that *nano* is often implicitly associated with something artificial rather than something occurring naturally.

In many cases, very small quantities go hand in hand with the high-throughput analog of a known technique. Beautiful examples can be found in the booming field of microfluidic devices and labs on chips. These new research platforms consist of silicon, glass or specialized plastic plates that are a few square inches in size and contain sophisticated channels, wells, reservoirs, pumps, valves, separators, and control structures in the micro- or nanometer range, which permit the targeted movement of very small amounts of fluids. Often handled by robots, they permit accurate spatial control, as well as the miniaturization, parallelization, and automation of biochemical or biological assays that, merely a decade ago, were done by hand in a Petri dish or with large machines on a laboratory bench. Intriguingly, these devices can be used as miniature assembly lines where cells or small organisms are subjected to sequential steps of complicated experiments while being moved through the channels of the platform. Because this sequence is computer controlled, the results are reliably precise and measurement errors are minimized.

An early milestone in the rush toward miniaturization was the implementation, in 1979, of a gas chromatograph on a small silicon wafer that could separate and quantify organic compounds. Shortly after came DNA chips and the incredible method of inkjet printing of cells into artificial tissues. The following decades witnessed the development of microvalves, micromixers, micropumps, and many other tiny devices that allowed the automated handling of complex liquid protocols. These rapid advances created such exuberance that proponents started to speculate about an impending revolution that would gain enormous power and make an impact similar to that of microelectronics a few decades before. After all, microelectronics was also the result of miniaturization, and amazing devices were made out of very small silicon chips with integrated electrical circuits. In the case of microfluidics and labs on chips, the jury on the revolution is still out.

Nowadays, microfluidic platforms are being used for a host of chemical and biochemical applications, from the synthesis, assembly, and breakdown of specific molecules to quite complex systems that allow the study of cells or small organisms. Because these micro-devices use so little material, tests are cheaper and results are often obtained more quickly than with traditional methods. It used to take several milliliters of blood for a diagnostic test. That is a humungous amount for microfluidic devices, where "miniaturization" refers to amazingly small quantities in micro-, nano-, pico-, femto-, and atto-liter ranges, which are all very, very small. Furthermore, microfluidic devices permit a high degree of parallelization, and pharmaceutical companies use them to perform hundreds or even millions of simultaneous assays to answer questions of dosing, toxicity, mutagenicity, and side effects of new pharmaceutical candidate compounds.

A beautiful example of the use of microfluidics in systems biology is the tiny roundworm *Caenorhabditis elegans*, which is an extraordinary organism. It is only about 1 mm long and very frequently found in soil, where it eats bacteria at a rate of up to 5,000 per minute. Maybe the most fascinating feature of the worm is that it always has exactly the same number of cells: 959 in the adult hermaphrodite, which has both female and male reproductive organs; and 1,031 in the rarer male. Scientists have been so fascinated by this discovery that they have figured out the exact developmental lineage of every cell from egg to adulthood.

C. elegans is quite a celebrity. Its 100 million base-pair long genome was one of the first to be sequenced among multicellular organisms, and detailed studies have characterized the function of many of its roughly 20,000 genes. Because of the constancy of the worm's cell numbers, it has been possible to map its "connectome"; that is, a more or less complete wiring diagram of its nervous system, which consists of exactly 302 neurons in the hermaphrodite. *C. elegans* has become so important for biomedical studies that an entire online database, *WormBase*, is dedicated to it. The worm has even spent some time on the International Space Station, was the subject of an International Worm Meeting in 2003, and led to three Nobel Prizes.

Fully automated microfluidic chips allow today's researchers to investigate a single worm in precisely controlled environments, under different chemical or physical stimuli, and even in a high-throughput manner. For example, squeezing the worm into a tapered channel on a microfluidic chip permits careful analyses of a variety of phenomena such as learning, memory, nerve regeneration, sleep, and the aging of its brain. *C. elegans* even serves as a model for nicotine dependence, because it exhibits the same stages of acute response, tolerance, withdrawal, and sensitization as humans.

C. elegans is translucent, which greatly simplifies microscopic imaging studies of its development and cellular differentiation. A remarkable advance for this type of imaging is the very recent introduction of the method of lattice light-sheet microscopy, which was used, among many other applications, for the subcellular localization and dynamic analysis of proteins during the embryonic development of *C. elegans*. In this procedure, an ultrathin light sheet sweeps through a specimen and excites fluorescence in successive planes, thereby allowing three-dimensional imaging of fast processes with very high resolution.

Molecular and systems biology, and medicine along with them, have been posing ever new and more challenging tasks, and the diverse fields of biomedical engineering, electrical engineering, and robotics have responded admirably, again and again. Usually, biologists suggested the original ideas and concepts and even developed the first prototypes, for instance in the early days of genomics and proteomics, and engineers took on these concepts and ideas and made them incomparably more efficient. This trend of collaboration has been extremely successful and will probably continue for a long time. In fact, the boundaries have become increasingly blurry between molecular biology and bioengineering.

If we look at the advancing developments of devices that systems biology uses on a regular basis, we might take it for granted that new devices will extract ever-more information from less material than had been thought possible merely a few years earlier. We may even become jaded by these successes. However, if we step back and look at a mass spectrometer that separates more than 30,000 chemical entities from a tiny sample of serum, or a microfluidic platform that allows us to study physiological systems in individual worms or even cells, we honestly cannot help but marvel at the combination of engineering and ingenuity and state in amazement: simply engenious!

6

Just a little bit

A bit became a bit in the 1930s and 1940s. In medieval England it had been a bite-sized morsel. In the young US, it amounted to an eighth of a Spanish silver dollar, and one could allegedly get a shave and a haircut for two bits. Then, in 1936, the distinguished American inventor Vannevar Bush, the lead founder of the early technology start-up Raytheon and head of the US Office of Scientific Research and Development during the Second World War, wrote a fascinating and quite amusing article reviewing the truly amazing mechanical and electrical devices that were being used at the time for all kinds of computing tasks, including differentiation, integration, correlation analysis, the solving of systems of algebraic and differential equations, the evaluation of wind tunnel experiments, and the analysis of numerous other diverse applications. While affirming that, "the combination of such machines with punched cards has made arithmetic into an entirely new affair,"[1] Bush lamented the size limitations of the cards and proposed that clever coding could increase their capacity to "over 300,000 bits of information."[2] About a decade later, the famous Bell Labs mathematician and statistician John Tukey, and the pioneer of information theory, Claude Shannon, formalized Bush's casual terminology by contracting *binary* and *digit* to describe a base unit

---

1 V. Bush (1936) Instrumental Analysis, *Bulletin of the American Mathematical Society*, 42(10), 652.
2 Ibid., 653.

of information as one bit, a flip-flop that could only take the values of 1 or 0, on or off.

Fast-forward to the twenty-first century, and there is no doubt that the little bit has conquered almost every bit of the world, and it is hard to think of any aspect of life that is not fundamentally affected by digitization. Building upon the visionary ideas of the nineteenth-century British inventor, engineer, mathematician, and philosopher Charles Babbage, who originated the concept of a programmable computer, ever-more sophisticated, powerful, and fast computers came to be, leading to today's huge supercomputers but, maybe more importantly, to huge numbers of very small computers that, in their own ways, are all supposed to make life simpler. Not surprisingly, computers have also become premier tools of systems biologists.

The most recognized and obvious help we get from computers is the handling of data. And whether it is the government, Walmart or systems biology, the amounts of some data have become so enormous that they are rightfully classified as BigData. In the context of modern biology, this aspect of computational data management is of immediate and enormous relevance in the field of –omics. In fact, large datasets from genomics, proteomics and metabolomics can often only be handled with methods of artificial intelligence and machine learning, as was discussed in an earlier chapter. Whereas the human mind is overwhelmed by complicated datasets, these computational methods are able to reveal association patterns, and these patterns are in turn the prerequisite for data interpretation and the further generation and formulation of hypotheses about the data.

While managing data is essential, computers do much more. Intriguingly in our context, they have changed how applied math is done and sometimes even dictate what type of math is feasible for a given task. Granted, paper and pencil are still the tools of the trade in many fields of mathematics, but they are increasingly being supplanted by entirely new approaches and strategies for solving problems. That the famous four-color theorem from 1852 was actually proven with a computer algorithm in 1976 probably made many a mathematician cringe. Vannevar Bush already forecast this development, arguing that devices, starting with the human hand, had always "influenced the course of formal mathematics,"[3] and that, after all, *calculus* is derived from the Latin

3 Ibid., 650.

word for pebbles that were used for simple arithmetic. Looking into the future, he predicted that,

> there is one influence which is undoubtedly coming, if equation-solvers become as fully developed and as rapid, reliable, and versatile as the arithmetical machines of commerce. Formal attention will be less directed to the mere solution of equations, in order that they may thus be rendered useful; and will hence be more directed to their formulation and interpretation.[4]

Computers can even solve problems that according to mathematics have no true solution. How is that possible? A good illustration and a very important example for systems biology is a set of differential equations. Most differential equations cannot be solved with methods of traditional mathematics. More specifically, we know that the solution of a typical differential equation is a function of time, but mathematical theory proves that one cannot explicitly derive this function. Unfazed by the theoretical impossibility, a modern computer algorithm easily computes "a" solution nevertheless. Admittedly, this solution is not the absolute truth, but it is close enough for essentially all practical purposes. What the computer does *not* do is present this solution as a true function of time, as a mathematician would like it. Instead, the computer generates numerical values of this unknown function at as many time points as we desire. Thus, we have solution points, but we cannot say whether these points belong to a particular function, such as a parabola or a logarithmic function. For most applications, this string of numbers is sufficient, even though it does not allow further formal analysis or mathematical proofs in the traditional sense.

To obtain the approximate solution, the computer uses a customized algorithm. With its name derived from the eighth-century Persian mathematician and astronomer Abu Ja'far Muhammad ibn Musa al-Khwarizmi, an algorithm is an iterative step-by-step procedure that grinds out each solution point with thousands of quite precise, yet approximate, calculations. As Bush put it, "The machine will not, of course, yield a formal result; it will give only approximate solutions." But on the positive side, he added very acutely,

4 Ibid., 666.

The machine does not care how complex the expression for a coefficient may be, so long as it may be plotted. ... Bizarre combinations [of functions] are exactly as readily provided for as is the usual case.[5]

Thus, we must forego some theoretical options of analysis, but the capacity of obtaining very good approximate solutions makes up for it by enormously expanding the domain of problems that can be addressed. Indeed, computational modeling in systems biology would hardly exist at all without this capacity.

The ease and efficiency with which differential equations can be solved computationally has rendered possible entirely new strategies for exploring biological systems. Most prominent among them are simulations. They fall into two major categories. One consists of targeted simulations of what-if scenarios. Typical questions are: What happens if the input (such as food) of the system is changed? Can the model, representing a cell or organism, tolerate large fluctuations in input? How often must medication be given to counteract the effects of a disease? What is the consequence of an enzyme that only has 40 percent of normal activity as a result of a mutated gene? If a model appropriately represents reality, simulations of such scenarios can answer uncounted questions. As just one application for the future, we will have health and disease simulators that permit the training of physicians and the determination of optimal treatment strategies.

Simulations in the second category are referred to as Monte Carlo simulations. They were invented by Stanislaw Ulam of the Los Alamos National Laboratory during the development of the first hydrogen bomb. The rationale for the new method was that the processes underlying a nuclear explosion are so complicated and contain so many parameters that nobody was able to predict with any reliability the possible or even likely outcomes of design changes. Thus, Ulam proposed a scheme for repeatedly sampling reasonable values for all parameters of the processes in a random fashion. For each random set, the equations were simulated, and the process was repeated thousands of times. All results were collected and then evaluated with statistical methods that identified the most likely outcomes, as well as best possible results and worst-case scenarios. In recognition of the random nature of the simulations, and

5 Ibid., 665.

dealing with such a somber application, Ulam euphemistically named the method after the capital of the Principality of Monaco on the French Riviera, which is home to many casinos. Monte Carlo simulations are only feasible if the model equations can be solved very effectively. Since that is usually a given in today's computing environment, the method has gained a lot of appeal for a vast variety of applications, including systems biology. Sometimes the simulations are derided as a brute-force approach, but they are effective and in many cases there is simply no viable substitute.

Another area of heavy use of algorithms in systems biology is optimization. A good example comes from metabolic engineering, where one is often interested in manipulating microorganisms into producing valuable organic compounds in modest quantities, or cheap compounds like ethanol in very large quantities. The overall strategy is to introduce targeted genetic alterations into the microorganisms that lead to changes in the yield of the desired compound. For a single function and for very small systems, it may be possible to compute optimal settings with traditional mathematical methods of calculus. However, this approach is no longer possible for systems of realistic size, especially if the optimal solution must additionally satisfy different types of constraints. For instance, concentrations and production rates of metabolites are limited by the physiology of the microorganisms, and exceeding them would compromise their viability. Thus, the optimization task consists of identifying model settings such that the product yield is optimal, while all constraints are satisfied.

Quite a different and very important optimization task in systems biology is the determination of parameter values that make a model match a given dataset as well as possible. This parameter estimation task may be approached with one of several conceptually distinct search algorithms. The oldest of these pursues the following approach. The user guesses a set of initial parameter values, and the algorithm computes the corresponding model outputs for this set and close-by sets. Because the initial estimate is guesswork and most likely not even close to optimal, it is lovingly called a guesstimate. Comparing the initial results suggests to the algorithm a direction in which changes in parameter values have a good chance of improving the current solutions. The algorithm selects new parameter values accordingly and evaluates the new solution. This two-step dance is repeated thousands of times, until no better solution seems possible. The situation is reminiscent of a hiker who is lost in the mountains. To make matters worse, thick fog allows him to see only a few feet in each direction. Even with this limited vision, it is possible to distinguish where the terrain goes up or down. Of course it could happen that the best path down the mountain first climbs over a ridge, but the best uninformed chance of returning to the valley is presumably to walk in the direction where the terrain slopes down. The hiker walks in that direction and after a while reevaluates the situation by again looking for the best direction. Search methods of this type are fittingly called "steepest descent" algorithms if the task is to find the minimum of a complicated function. For finding a maximum, essentially the same method is called a "hill climbing" algorithm.

Descending and climbing algorithms often make slow progress, especially if there are many locations that are not optimal but still better – that is, lower – than their surroundings, such as a mountain lake. These almost optimal locations are treacherous, because algorithms tend to get stuck in them: the algorithm looks around, doesn't find better solutions, and declares victory, even though it did not find the truly optimal solution. Greatly confounding this situation is the fact that the true optimum is of course not known, so that the user can never be sure whether the algorithm truly succeeded.

In their search for genuine alternatives with better performance, mathematicians and computer scientists looked for inspiration from nature. A particularly promising resource was evolution, due to the strong belief that evolution repeatedly uses small alterations and then selects the fittest, thereby weeding out the weak and promoting the strong. As a result, organisms and new species are thought to become better and better over time. Trying to mimic this successful, stepwise optimization, an entirely new class of evolutionary algorithms emerged. Similar to the earlier search approaches, these algorithms depend on millions of small system evaluations, but each improvement step is achieved in a distinctly different, biologically inspired fashion.

The most prominent example is the genetic algorithm. Suppose the task is to find a set of parameter values that renders an optimal model fit to the data of interest. Somewhat simplistically, although not too much, the genetic algorithm strings all the parameters together in a chromosome, so that the first section belongs to the first parameter, the next section to the second parameter, and so on. To start the algorithm, a few dozen different chromosomes are created by filling the sections with guesstimates for all parameters. Now, pairs of chromosomes are allowed to mate. As in genetics, there is crossover, so that part of the offspring is from mom's chromosome and the rest comes from dad's. Mutations are allowed at a very low rate. The result of each mating event is a chromosome that contains a new combination of parameter values. All these combinations are run through the model, and if any of them yield better results than earlier chromosomes, they are kept, while the weakest parent and offspring chromosomes are discarded. In other words, the fate of all chromosomes depends on their quality, or fitness, which is measured as the closeness with which the model results match the data. The very weakest chromosomes usually disappear, while the others have a higher or lower chance of mating in the next round, depending on their fitness. Miraculously, iterating through a large number of mating events often leads to very good solutions. These are usually not absolutely optimal, but close.

It seems that different variations on the theme of biologically inspired algorithms are beginning to take over the animal kingdom, because there are now bee and ant colony algorithms, swarming and flocking algorithms, cockroach and bat algorithms, even bacteriologic algorithms, and a number of others that are all reminiscent of the actions of critters who have survived evolution and become very good at finding food or completing other tasks.

Applications of optimization abound in numerous other areas of biology. One of the fundamental tasks in bioinformatics is the comparison of genes from different organisms or species. Because each gene has been affected through the eons by mutations, deletions, insertions, and rearrangements, such comparisons are complicated. Many algorithms have been developed for aligning two genes such that their bases match as well as possible. Another application of finding optimal situations is molecular modeling. A standard task here is to predict how a string of amino acids will fold into a three-dimensional protein. All amino acids have chemical and physical properties that attract or repel parts of other amino acids, and a feasible configuration of the protein is one where the total energy of the molecule is low. While the computational task is conceptually easy to grasp, even the most advanced algorithms have a hard time finding the configuration with minimal energy. A related question is the determination of where on the surface of a protein a smaller molecule, such as a prescription drug, will bind.

Quite a different task for computer algorithms in systems biology is the standardization of its language. While Juliet may have told Romeo that "a rose by any other name would smell as sweet," computers do not like ambiguity, and crisply defined names and terms are very important when it comes to automatically analyzing texts or databases. A first and relatively easy aspect of this task is to establish a list of all synonyms of a biological entity. Much more difficult is a characterization of its specific role. For instance, a protein might be responsible for the transport of ions in and out of a cell. This function is a feature to be coded and associated with the record representing the protein, along with its typical location, corresponding gene sequence, and other aspects. The modern sub-field of systems biology addressing these issues is ontology research. Originally a fundamental topic of philosophy that addresses the nature of being and of the existence of things, ontologies have become the foundation for effective methods of computationally reading and interpreting databases and the texts of scientific publications. With a strong ontology, it is even possible to perform computer-based reasoning that can suggest the functionality of a gene or protein, based on its associated information.

In the last paragraph of his 1936 article, Vannevar Bush said,

> We are in an age of complex instruments. Out of it will come devices that will revolutionize the use of mathematics, and will profoundly influence some branches of mathematics itself. This process is now beginning, and it is probable that the next decade will see important advances.[6]

That paragraph still rings true today, and as biology becomes increasingly quantitative, it will become ever-more relevant. The little bit has come of age and many great advances in systems biology will depend on it.

6 Ibid., 666.

7

Supermodels

Models come in many forms. We all use conceptual models on a daily basis. Driving to the store, we know when to turn right or left, because we have in our minds a mental picture of the street scenes. Even if we do not know a city well, we have learned how to read and interpret a street map, which in itself is a model of the city. Dolls and toy trains are physical models that allow children to learn much about the world of adults. A good architect sees from a blueprint what a building will look like. In addition to conceptual models, which always whirr about in a scientist's mind, systems biology makes heavy use of mathematical and computational models. The difference between the two is actually quite vague, as many mathematical models are analyzed with computers and computational models are based on mathematical formulae and equations.

A typical model in systems biology consists of a mathematical description of processes occurring in cells, organisms, populations or ecosystems. To see why such models can be helpful, consider, as an analogy, the computer system in an airplane as it prepares for landing. It takes input information from the real world, such as the speed and weight of the plane, power of the jets, current altitude, length of the runway, as well as environmental conditions like wind speed and direction, enters all these data into a large system of mathematical equations, evaluates these equations, and determines the appropriate settings of rudders, flaps, and slats that ensure a smooth landing. The concepts in biology are similar, and one might imagine inputs regarding the health status of a diseased person, which are computationally converted into suggestions for a treatment. The modeling process itself is more complicated than for engineered systems because we often do not know the biological component parts and processes sufficiently well. It also turns out that knowledge of the parts is not sufficient to reconstruct a biological system, as we will discuss later.

In his 1858 book, *The Autocrat of the Breakfast Table*, the American physician, writer and poet Oliver Wendell Holmes mused, "I find that the great thing in this world is not so much where we stand as in what direction we are moving." Maybe Mr. Holmes should be made an honorary modeler, because his statement is a perfect layman's description of an ordinary differential equation, which is the bread and butter of modeling dynamical systems in biology. An ordinary differential equation does not so much define the present state, size or extent of a system, but rather describes in mathematical terms in which direction the system is moving. As an example, the size of a population is determined by processes that add individuals by birth or immigration and processes that reduce it. Any changes in the total number are governed by these processes. A similar situation is true for metabolites that are synthesized and degraded, as well as a host of other phenomena. A differential equation formalizes these functional relationships between the processes that occur within a system and the temporal changes they effect in the state of the system as a whole.

The concept of relating overall changes to individual processes down in the trenches is worth pondering. Importantly, this relationship lends itself very well to biological experimentation, because overall changes are often directly observable, and the details of processes, studied one at a time, can be assessed with experiments more readily than the entire state of a system while it is undergoing changes. If implemented appropriately, the formulation of a differential equation system also automatically accounts for synergisms and other complex interactions, which often are difficult to assess otherwise. Once a differential equation is formulated, it secondarily permits the computation of the state of the system at any desired time points in the future. It can thus be used for explanations, prognoses, and predictions of future events.

Mathematical models of biological systems have by and large co-evolved with the data that were available at the time. One of the early successes was the formulation of so-called *mass action law*, with which the Norwegian mathematician Cato Guldberg and his colleague, the chemist Peter Waage, described the kinetics of elementary chemical reactions. In their seminal paper of 1864, they assess the affinity, or chemical force, between two reacting substances as proportional to their amounts, [A] and [B], each raised to a particular power, which depends solely on the nature of each substance. Inconveniently, the "particular powers" made the result mathematically complicated and in retrospect appeared to be unnecessary for the data they had, motivating Guldberg and Waage to eliminate the powers altogether in their next publications. Intriguingly, 100 years later, the generic modeling framework of Biochemical Systems Theory reintroduced the same powers as very useful parameters that permit great flexibility not just for kinetic reactions but also in models of a host of biological situations in physiology, ecology, and epidemiology.

This short blip in history provides a hint of how mathematical models were designed in the early days of biology: a repeated observation led to an idea for a hypothesis, which was formulated in the language of mathematics. Sometimes it had to be simplified due to computational difficulties. The approximations resulting from these simplifications were justified with mechanistic explanations and hopefully supported by experimental data.

A noteworthy situation illustrating this chain of events occurred a few decades after Guldberg and Waage, at the beginning of the twentieth century. Biochemists at the time had become very interested in reactions that are only possible with the assistance of enzymes. They investigated these reactions by experimentally measuring the rate at which a substrate is converted into a product. In a wide range of cases, the biochemists found a similar, increasing trend between substrate amount and reaction speed: the more substrate, the faster the reaction. However, as we would expect from the law of diminishing returns, adding more and more substrate eventually did not increase the speed any further. The repeated observation of the same saturating relationship for different reactions suggested that the process was apparently governed by a clear, quantifiable functional relationship. Intrigued by the challenge of establishing such a mathematical relationship, the French-Russian physical chemist Victor Henri, and later the German biochemist Leonor Michaelis and his Canadian student Maud Menten, proposed a molecular mechanism according to which the substrate first forms a chemical complex with the enzyme. Subsequently, they proposed, this complex would break apart and in the process release the product of the reaction as well as the enzyme, which could be used over and over. To make their insights quantitative, Henri, Michaelis, and Menten designed a mathematical model describing the proposed mechanisms, which they formulated in terms of the earlier mass action equations of Guldberg and Waage. The result consisted of a system of three differential equations which, however, could not be solved. To overcome the problem, they made some clever assumptions and simplifications, which they considered as likely true in many relevant situations, although not always. These simplifications reduced the system of differential equations to a very simple, explicit function relating the reaction speed to substrate and enzyme availability. The result was an approximation, but this approximation turned out to be very powerful and sufficient for most practical purposes. Indeed, the mass action equations and the rate function proposed by Henri, Michaelis, and Menten are still in regular use today, together with uncounted variations on the same theme, and tens of thousands of papers have made use of these old concepts.

The early models have been very useful for the analysis of single reactions. However, as soon as a dozen or a few hundred reactions need to be considered simultaneously, they become cumbersome. One might argue that we have big computers nowadays so that "cumbersome" is no longer a valid criterion. This argument is true in the sense that we can easily evaluate hundreds or thousands of equations. It has been demonstrated quite clearly, though, that the mathematical structure of these formulations makes insights into the hidden features and complexities of large biological systems very difficult, if not impossible.

Thus, to accommodate today's world of rich –omics data, modeling must employ mathematical and computational models that can readily be scaled up to realistically sized systems. The quaint "sandbox" models of the past are to be left behind. What we need are supermodels! Models that are able to tackle the complexities we are encountering in large biological systems, which allow us to study static and dynamic features, and that ideally permit extrapolations to untested conditions and make correct, reliable predictions. And the supermodels should do all this quasi with a smile; that is, in a fashion that is mathematically tractable and can be computationally executed on a PC with great efficiency.

Intriguingly, nature has not provided us with guidelines for how to set up such models, and the choices are *a priori* infinite. As a consequence, many modelers have used ideas from physics, made more or less reasonable assumptions regarding their model types, or just looked for models that happened to fit the data. The problem with many of these models is, again, that they are difficult to scale up to large sizes and become unwieldy. An interesting alternative to this piecemeal approach are so-called canonical models, which use the same types of mathematical function for all processes in a system. This feature may sound overly restrictive, but experience over the years has demonstrated that canonical models are mathematically well-justified, surprisingly accurate, have numerous beneficial features for the design, diagnosis, and analysis of models, and can be scaled to larger problems with relative ease.

For computational systems biologists, the mathematical details of the various modeling approaches are central to their research. Here, however, it appears to be more important and illustrative to discuss what types of model are available and what their genuine added benefits might be.

In terms of types, differential equation models are the undisputed winners when it comes to frequency of use, but they are not alone. There are many models with rather different structures, such as those that assess the probability of something happening; one might think of a cell turning into a tumor cell. Models may also consist of agents and rules and resemble sophisticated computer games. Each agent may be a person, a cell, a molecule, or some other biological entity, and the rules determine what each agent is able to do and what happens if two agents meet. Initially used in the area of consumer behavior, these models are being used increasingly in systems biology. The rules and agents can be formulated with mathematical rigor in unlimited variations, and agent-based models have become interesting research tools. As an example, one might use such a model to study the processes during the development of a fertilized egg cell into a fully formed embryo. Agents could be cells of different types in their proper locations, while the rules would govern cell migration, differentiation, and development. Agent-based models can even help us analyze emergent responses of systems, such as the onset of oscillations. We will discuss this property in Chapter 9.

No matter what type is chosen, mathematical models are formulated in their own language, which is a mixture of our common language with mathematical symbols, terminologies, and methods of crisply defined logic. This language encapsulates mathematical knowledge and theorems that have been proven true over the centuries and are used to make valid deductions regarding the features and outputs of a model. Importantly, this language permits the rigorous and unambiguous definition of functions, equations or rules. These in turn contain variables, which often represent the quantities of greatest interest in a model analysis and typically change over time. A variable could be the size of a growing population or the glucose concentration in blood. The sky is really the limit. Models also include parameters, which characterize the specific quantitative features or settings of a model. They are fixed for a given computational experiment, but may be different in the next experiment. A parameter might simply be the surrounding temperature. It could also represent the number of receptors on the surface of a cell or describe the efficiency with which an enzyme catalyzes a reaction.

A key ingredient of any modeling effort is abstraction: not every detail of the real system can or should be considered in the model, lest the model become overwhelmingly complicated. A study of the flight of butterflies will probably ignore the coloration patterns of their wings. This abstraction process, along with assumptions and decisions regarding what should or should not be included in the model, and how it should be represented, seems rather innocuous. However, it is the most critical step in modeling. It is also the step that is most prone to criticism by experimentalists, because they may not agree with the decisions or assumptions the modeler made. Once the abstraction is accomplished, the rest of the modeling effort should really be beyond reproach, because it strictly consists of the application of tried and true mathematical and computational techniques, which the modeler is expected to execute correctly.

The biological reality and the realm of the mathematical model are separated by a gap, which is bridged by correspondence rules. In one direction, these include abstractions, omissions, and the definition of the mathematical quantities. For instance, one might call all metabolites M with a subscript: M_j. M_1 could represent glucose, M_2 lactose, and so on. Subscripts assure that the modeler never runs out of names. The correspondence rules record these settings. In the reverse direction, the correspondence rules explain what the mathematical findings mean in the language of biology. For instance, the correspondence rules must "translate" what "a system has a Hopf bifurcation for p at 1" means: the system was residing at some state, but all of a sudden begins to oscillate when one changes the value of parameter p from less than 1 to greater than 1. Thus, the correspondence rules transport the most important items from the biological reality into the realm of math. The computational analysis occurs entirely within this realm, and then the results are interpreted and transported back by the correspondence rules to reality, where one gains new biological insights and a deeper understanding.

The generic modeling process may be divided into three phases. The first consists of the choice of model type, such as a system of differential equations or an agent-based model. This choice is governed by the crucial triad of crisp research questions to be answered by the model, sufficient data, and a suitable model structure. The first phase also includes the design of the specific model and the determination of parameter values. The second phase is dedicated to diagnostics and refinements. It is purely technical and tests the model against data and general expectations. For instance, one should expect a natural system to be robust and tolerate modest alterations in inputs or external conditions. Indeed, if a mathematical model is very sensitive to small perturbations, something is probably wrong. This second phase is usually time-consuming, often taking many months, because one frequently detects unanticipated discrepancies, and these must be carefully analyzed and corrected through changes in parameter values, refinements in the model structure, or the inclusion of components and processes that had been omitted. The third phase employs the model to answer the research questions that were posed at the beginning.

A central question then is: what exactly are models in systems biology supposed to show or do? Indeed, they may have quite different roles. Prominent among them are bookkeeping, predictions, explanations, explorations of possible system responses, guidance for future experimental manipulations, and help with the formulation of novel hypotheses.

As we discussed before, –omics datasets are often so large that it is impossible to discern association patterns with unaided intuition. Mathematical and computational models are perfectly suited for this task of organizing data and extracting complex patterns from large datasets, because they don't forget details, even if these are in the thousands. Models can also separate important information from background noise and distinguish signals from natural variability, thereby reducing the complexity of the dataset. Furthermore, by being able to account for all relevant data in their appropriate contexts, the models allow us to organize our thinking about the system and can assist in the stepwise formulation and screening of hypotheses.

One role of a model, which is often considered its greatest asset, is the power of prediction. The concept is simple. We formulate, diagnose, test, and analyze a model, and once we are convinced that its performance is adequate, the model should be able to tell us something new. What would happen if we increased the input to the model two-fold? How would a disease system respond if we administered a specific drug? Will a yeast culture produce more alcohol if we increase the temperature? These types of prediction are similar to the weather forecast. The National Weather Service operates huge weather models, which have been programmed to account for all major processes that can possibly affect the weather. The models receive yesterday's and today's specific weather data and, based on this information, are expected to predict if the weekend will be sunny. In the future of systems biology, we will likely have similarly huge models that are programmed to account for all physiological processes related to the heart or the liver. We will supply the computer with all the data measured in a particular patient, and the model will make a prognosis about the likelihood of the patient having a heart attack within the near future. Alas, the acquisition of all necessary data in biology is much more difficult than in meteorology, and comprehensive predictions in systems biology will not be as reliable as in weather forecasting for quite some while. Similar to weather forecasts, the disease models of the future will predict likelihoods, rather than absolute truths, because neither we nor the computer can know all the processes that could possibly affect the outcome.

Closely related to a model's power to predict is its ability to explain. If we were able to rationalize why and how a specific gene is functionally connected to a particular disease through RNAs, proteins, signaling processes, and metabolic changes, we could potentially exploit this information to prevent, cure, or at least ameliorate the disease. Model-based explanations are arguably of the greatest value if a model response is counterintuitive. For instance, it happens that increasing the input to a system a little bit increases some important response, but that increasing the input more strongly actually leads to a decrease in the response. Internal regulation, the turning on or off of subsystems, synergisms and antagonisms, as well as various threshold effects may cause this phenomenon, and it takes a strong dynamic model to quantify it appropriately.

Models also allow us to ask more generally what the possible repertoire of responses to external stimuli could be. For instance, one might ask whether a particular variable could possibly exceed some threshold or if a population is likely heading toward extinction. Analyses of this type are possible with automated simulations that scan very many combinations of values for many of the system's parameter values. Each simulation is performed very quickly, but these types of scan may require millions of iterations. In the end, one obtains answers to the above questions or statements that quantify how likely or unlikely specific outcomes of interest are.

It is human nature to want to change things to our advantage. A good example in the field of biological systems modeling is metabolic engineering. The goal is to manipulate microorganisms so that they produce an organic compound that is of industrial interest. A specific goal may be to "reprogram" the bacterium *E. coli* to make chemical precursors for prescription drugs or to alter bacteria or yeasts so that they produce copious amounts of biofuel from inedible plant parts like wood chips and corn stover.

A typical illustration is the industrial production of citric acid, which is used all around the world as a preservative for foods and soft drinks, as well as a starting product for chemical syntheses and as a cleaning agent. The saga started almost 100 years ago, when James Currie of the US Department of Agriculture observed that a large number of strains of the black fungus *Aspergillus niger* very actively produced and excreted citric acid. He modified the medium and selected for high producers, mostly through trial and error, up to a point where the United States became self-sufficient in the production of citric acid and even exported some to Europe. His work led to several hundred patents. Some modern metabolic engineers attempt to speed up this type of process by eliminating the need for trial and error. The generic approach consists of measuring with both traditional and –omics methods as many aspects as are needed, merging all this information into mathematical models, optimizing these models by altering gene expression or enzyme activities, and implementing the optimized gene and enzyme profile in an actual microbial culture. Clearly, the better the models are, the more reliable the predicted increases in productivity will be.

In contrast to common wisdom, an answer in math is not always either right or wrong. Especially in the field of modeling, no model is entirely correct, due to simplifications and abstractions, but many different models may be sufficiently close to correct to be of practical use. The quality criterion for a model structure thus switches from true-or-false to better-or-worse, where the latter not only accounts for the quality of matching data but also for insights that may be gained and for the time it takes to obtain answers.

One might think that better models always require a large number of variables and a high degree of complexity, but that is not necessarily true. In fact there seems to be no clear correlation between the complexity of a model and its value. For instance, very simple models are often used in forestry when the complicated decision has to be made either to cut a tree stand or to wait another year in order to let biomass increase. The simple models occasionally fail, of course, but they are usually much more robust and reliable than very comprehensive, physiological models. In the end, the distinction of the "best" model hinges on one criterion: how well and reliably does the model answer the questions posed in the beginning? Modeling the heart as an adaptive, peristaltic pump might be the best approach to understanding the physical forces that are stretching and straining the heart muscle throughout a lifetime. But such a model will clearly not be able to characterize the role of a gene that had been associated with an increased risk for myocardial infarctions. In the end, the model offering the most insight is to be preferred, and because data, as well as questions and methods, change over time, today's best models might not be the winners 20 years from now. It is not even necessary that a good model matches all observations perfectly. Sometimes a simple model that unambiguously reflects the correct trend in the data provides more insights and is therefore the best choice.

The discussion in the previous paragraphs leads us to the question of what types of issue the systems biology models of the future are supposed to address. As is often the case, the answer depends on who is being asked. One Holy Grail of the field is a set of reliable, comprehensive supermodels of whole cells, organisms or diseases. Such models would be enormously helpful, for instance, in pharmacology and prescription drug testing. Instead of having to perform hundreds of animal experiments, the drug could be administered virtually to the computer model, which would quickly show its main effects as well as its side effects. One would not necessarily rely on these models alone, but if the model convincingly demonstrated unanticipated side effects, the drug candidate would not be considered further, and millions of dollars potentially spent on clinical trials could be saved. The potential of reliable whole-cell or whole-organism models is clearly unlimited.

There is a second Holy Grail, namely, the successful creation of models that reveal a true understanding of the laws and principles that govern biological systems, as we discussed in Chapter 4. The typical question to be answered is very different from the questions we encountered before, namely: Why did evolution result in a structure as we observe it and not in a different structure? What is the advantage of a particular design, such as the shape of a shark fin? Why is feedback inhibition so prevalent in biology? What is the best control strategy for a given process? In simple cases one might be able to speculate about the rationale behind a design with intuition, but in more complicated cases this is no longer possible, and one needs to develop appropriate models to answer such questions.

The characterization of general design and operating principles of biological systems is important for two reasons. First, it allows us to understand new systems more quickly; second, a good understanding of these principles helps us avoid bad mistakes when we start manipulating or even creating organisms for specific purposes, as is the goal of the new field of synthetic biology. If the principles are very general, they may even be considered "laws of biology," of which we have very few so far. These laws could eventually become theories of biology, and the associated models would be true supermodels that would allow us to make precise and reliable predictions regarding many untested situations. A theory is very potent, as we can see in physics, and it is not just an esoteric construct but has uncounted practical applications, as we can see in engineering, which makes practical use of the laws and theories of physics. As the Prussian psychologist and philosopher of science Kurt Lewin once said, there is nothing more practical than a good theory, and that is without doubt true for systems biology.

8

Close only counts in horseshoes and hand grenades

It was allegedly major league baseball manager Frank Robinson who inspired the phrase, emphasizing that "close don't count in baseball." Only one team wins the World Series in a given year, even if the point spread in Game 7 is just one. Indeed, "close" is unacceptable for many aspects of life. Proper bookkeeping does not tolerate missing dollars or even cents. Almost winning the lottery, or almost getting the dream job, let alone the dream girl or guy, simply isn't good enough. Sometimes, "almost" seems even worse than missing the target by a wide margin.

If life with all its vagaries and uncertainties is so often dissatisfied with "close" or "almost," one would probably expect that hard-core science and engineering are even tougher when we miss the mark, even if only barely. And what does mathematics, the most precise of all human endeavors, have to say about "close enough"? It may come as a surprise, but science, engineering, and mathematics all embrace the concept of only being sufficiently close rather than 100 percent accurate, as long as the deviations are handled appropriately.

Two reasons make the quest for uncompromising precision infeasible, especially in a field like systems biology. First, we seldom know what the exact and precise truth is, and nature does not come with an instructions manual that offers guidance regarding the choice of perfect models. Second, the truth is usually too complicated for us to comprehend in its entirety, let alone convert into a computational representation. As an illustration, suppose we are interested in constructing a computational model of a cellular signaling process. With a coarse-grained perspective, the task is not all that difficult. The cell receives a signal in the form of a physical change in the environment, such as an electrical or mechanical impulse, or a chemical stimulus, such as the arrival of a hormone that was sent from a different location in the body, and responds by synthesizing the proteins or metabolites it needs. We could model this signaling mechanism like a light switch that turns the appropriate processes on or off. In mathematical terms, a simple toggle switch between 0 and 1 would do fine, and for some purposes, such a model might indeed be sufficient.

However, if we want to understand signal transduction more deeply, or if something goes wrong with the signaling mechanism, thereby causing disease or other undesired consequences, we need to study the process in finer detail, and the situation immediately becomes very complicated. Typically, the signaling molecule attaches to a receptor on the outside of the cell surface. This receptor is a protein that is embedded in the cell membrane, with one end sticking out to bind the signaling molecule and the other end dangling into the cytosol inside the cell. The binding of the signaling molecule causes the receptor protein to change the shape of its internal end. The shape change in turn triggers a series of phosphorylation events, which are mediated by a specific cascade of enzymes that are organized along protein scaffolds. This enzyme cascade amplifies the signal and in the end triggers a response. For instance, it might cause a specific transcription factor protein to move through the viscous cytosol to the nuclear membrane, cross this membrane, and enter the nucleus. Once in the nucleus, the transcription factor finds matching stretches of DNA and causes changes in gene expression. These changes in turn trigger the production of mRNAs, which ultimately lead to the synthesis of target proteins that are responsible for the appropriate response. Such a response typically consists of the production of specifically needed metabolites or further signaling or structural proteins. Not to forget, this chain of processes requires energy, ribosomes, nucleotides, amino acids, and other molecules, along with a host of secondary factors. All of these events must be correctly organized within the three-dimensional, very crowded space of the cell. Of course, all component steps obey the laws of physics, but each step is so complicated, and the processes collectively are so convoluted, that it is simply impossible to write down a comprehensive, mathematical representation that captures the physical features of the signal transduction process in its entirety.

Our inability to capture the details of fundamental biological processes like signal transduction in a computational model is quite interesting, as it stands in stark contrast to our mathematical and computational prowess when it comes to elementary physics and engineering. There, the situation is different, because we have strict laws that apply again and again, without fail, at least within our macroscopic world. We know how to describe mechanical forces or an electric circuit with mathematical rigor, and these descriptions are correct today, as they were in the past and will be in the future. Furthermore, we can combine these formulations into descriptions of very complicated machines, test them, and make predictions that we know are true. Just imagine the huge number of processes that govern a space mission, like the exploration of Mars with the rovers *Spirit* and *Opportunity*. Every step along the way, from launch to travel to landing, involved thousands of mechanical and electronic parts whose roles and functions were captured accurately in mathematical models and computer programs. In the end, as predicted and hoped, although never before tested, the rovers landed and began to roam the surface of Mars, exploring its physical and chemical features, and beaming awe-inspiring pictures back to Earth.

If we are able to capture these complicated processes mathematically with such precision and reliability, why can't we do the same in systems biology? The dominant reason is the following. In addition to relying on laws that govern elementary physical processes, engineers have become very good at designing gadgets and their components to precisely prescribed specifications. We can construct engines that reliably do exactly what they were designed to do. Using control mechanisms, we can furthermore make sure that key components only receive inputs that we permit and prescribe, and if such inputs are presented, we know exactly how a gadget will respond, because we know its design features. By being certain of all input–output relationships, and by employing designs that adhere to the principle of superposition, it is feasible to assemble ever larger and more complicated machines while still being able to make very reliable predictions regarding their responses to different inputs.

Contrast this situation to biology. First and foremost, biological components are what they are. We did not design them, we sometimes do not know their features or even of their existence, and we often do not know their specific roles and the characteristics of their interactions with other components. As a case in point, there are still many human genes and proteins whose function we simply do not know, in spite of intense research over the past decades. Instead, most of our knowledge about biological systems is gleaned from observations of their outputs or their responses to perturbations, and it is our task to infer from these observations how the systems are internally structured and how they function. This inference task is often very difficult. To make matters worse, the biological components often function differently, or not at all, if we take them out of their natural milieu for testing. Moreover, many processes and components act in concert with each other, and their collective behavior is the result of a complex conglomerate of physical functions that is very difficult to tease apart.

Interestingly, the outward responses of biological systems are often rather simple, even if the systems themselves are complicated. Take, for example, a bacterial culture. If we supply the bacteria with adequate nourishment and conditions, the culture starts to grow, and if we plot the population size against time, we often obtain a simple, smooth S-shaped growth curve that starts with a small number, picks up speed and then flattens out. A single function or differential equation easily describes this growth trend very well. The simplicity of the trend is not merely due to the fact that we are looking at bacteria. Pediatricians regularly use size charts for girls and boys, and the growth trends in these charts have surprisingly narrow bounds. In other words, most children of a certain age are within a few inches of height, and predictions regarding their future growth are much more reliable if they are based on these simple trend lines than if one tried to derive them from detailed physiological measurements.

The outward simplicity of larger systems tends to be a matter of statistics. As an analogy, consider an ideal gas in a closed container. Gazillions of atoms are whizzing around, bumping into the container wall and colliding with other atoms, and it is impossible to track or compute the exact path of any single atom. Yet, the relationship between the volume, pressure, and temperature of the same gas as a whole follows a simple and very reliable law. This law is effective, because it is a law of statistics, which tends to become more and more predictable as the number of involved items increases. A bacterial culture or a growing human is of course much more complicated than a gas, but it is the statistics of many molecules and cells that keeps a normal growth process within surprisingly narrow bounds. As the flipside of this argument, these laws become less reliable if fewer components are involved, and we can see that they are really approximations that are consistently accurate only at the right scale.

Thus, we are faced with an intriguing situation, which almost sounds like a paradox. Namely, at a fundamental level, biological processes are governed by physics, which we usually understand quite well, and input–output relationships at this level are easy to represent with mathematical models. At a very high level, biological systems often exhibit rather simple responses, which are also easy to represent. Yet in between, the inner workings of life are very complicated, and even apparently simple systems are difficult to comprehend and can be full of surprises. Alas, it is at this intermediate level where systems usually break, where engineering promises the greatest gains, and where therefore the most interesting investigations in systems biology should be performed. And it is at this level that we encounter the biggest challenges formulating appropriate models.

There are no silver bullet solutions, let alone strict laws that would help us approach these complex phenomena in the intermediate range. However, a very important component of nearly any imaginable computational solution is the concept of an approximation. Roughly speaking, an approximation consists of replacing a complicated function with a simpler function. As a practical illustration, we could deem a tree trunk as straight, even though it truly is not, if we look closely enough to see all the little indentations and protrusions in its bark. But from some distance, the trunk looks smooth and straight. An immediate gut reaction will be: is such a simplification allowed in an exact science or in math? As so often, the answer is "it depends." The most obvious question is whether the deviations of the tree trunk from absolute straightness are important to us. If they are, then the approximation will eventually fail us. By contrast, if we are in the lumber business and the trunk is straight enough to yield good boards, the declaration of straightness may just be sufficient.

What is of real importance in systems biology is that an approximation, such as the replacement of a nonlinear function with a linear function, has a much simpler mathematical structure than the true function, which we often do not even know. This simpler structure frequently allows types of analyses that would be much more difficult, if not impossible, for the original functions. As a representative example, consider linear regression, which can be performed even with a modestly sophisticated pocket calculator, whereas nonlinear regression is in many practical cases so complicated that even large computers experience difficulties.

Importantly, and perhaps surprisingly, we can compute approximations even if we do not know the true nature of the approximated function. As soon as we declare that we choose to use a linear approximation, we can compute all of its features, such as its slope and intercept, independent of the mathematical features of the approximated, unknown function.

Linear functions are very special in mathematics, and no other subfield offers as many tools and solution strategies. Nonetheless, they are not the only choices for approximations, and parabolas, exponential functions, logarithms, and power-law functions extend our repertoire toward nonlinear functions that often capture biological phenomena more accurately than linear functions, yet have features that simplify all subsequent mathematical and computational analyses. This enhanced freedom is very welcome, but comes with a new and intriguing challenge; namely, which approximation to select in a given situation. At this point there is no agreement on this issue in the field of systems biology. Indeed, there is not one optimal formula for any given phenomenon, and different laboratories use different functional representations, based on the relative importance they place on mechanistic arguments for one formulation as opposed to another, mathematical and computational tractability, willingness to accept smaller or larger residual errors and, perhaps hard to believe but true, personal taste. As a consequence, a biological phenomenon may legitimately be analyzed with very different model structures. Even if the same model structure is selected, it is possible that different parameter sets offer similarly good representations of the phenomenon, an observation that has been termed sloppiness. Thus, surprisingly, one can legitimately be sloppy even in mathematics.

No matter which approximation we choose, we have to admit that it is not "the real thing," which sounds better than saying that it is "wrong." Yes, there are no free lunches, not even in mathematics. By using an approximation we buy simplicity at the cost of residual errors. We may or may not be able to quantify these errors, but they are there and cannot be argued away. So, what our teachers told us, that in math it's either right or wrong, is strictly true, but this black-and-white statement does not convey the whole story, and the real dichotomy in systems biology is often, rather, whether one approximate model is more useful than another. In other words, when we apply math to the real world of biology, we are forced to compromise. We must weigh simplicity and analytical and computational tractability against residual errors we are willing to accept.

Before we get depressed about unavoidable approximation errors, we should return to our role model: physics. Physics appears to be so much more rigorous and clean than biology that it seems immune to discussions of good and bad models. However, if we inspect it very closely, we can actually see some cracks in our perception. As an example, let's consider gravity. Our earthly experiences suggest that the law of gravity is ubiquitous, unambiguous, and timeless. Yet, this fundamental law is no longer exactly true when we venture into the far reaches of the universe, where the law gives way to the principles of relativity theory. Nor does the law of gravity hold at the other end of the spectrum, namely, at the scale of quantum physics, or even nanoparticles, where gravity is overshadowed by other forces that we do not perceive as influential at the macroscopic level. These sobering facts force us to conclude that one of the most fundamental laws of physics is "only" an approximation in disguise, even though this approximation is excellent at our human scale of perception. The same is true for most other laws in nature. They are approximations if pushed to the extremes. But if the pervasive use of approximations even in physics is still no consolation, maybe Aristotle can put us at ease. He allegedly proclaimed, "It is the mark of an instructed mind to rest satisfied with the degree of precision which the nature of the subject admits and not to seek exactness when only an approximation of the truth is possible."

Systems biology has actually more direct justification to embrace approximations than Aristotle's wisdom. Nature itself lends its support to their validity in fundamental features of living systems. One prominent characteristic of many biological systems is directly related to the fact that approximations are great, or at least acceptable, if they are applied within some restricted domain, although maybe not outside this domain. Even the straightest tree trunk eventually branches out. More generically, nature uses multitudes of control mechanisms to keep important processes and systems within small operating ranges, so that approximations are actually adequate over these ranges. Our body keeps its internal temperature almost constant, and we start being concerned about fever or hypothermia if the temperature is merely five or six degrees outside the norm, while we at the same time can tolerate outside temperatures within a very wide range. The physician's desk reference indicates that electrolytes and other components of our blood serum must have concentrations within narrow ranges, otherwise we are declared sick. Generally, many physiological and cellular processes are buffered such that molecular key components are kept within surprisingly small operating ranges most of the time. Within these ranges, it appears, approximate models may just be sufficient.

Another fundamental feature that supports the use of approximations in systems biology is called "separation of time scales." Our life span in the twenty-first century is maybe 80 to 120 years. Such a time period is incredibly long in comparison to most chemical and biochemical processes that occur in our cells every day with speeds at the order of fractions of seconds. At the other end of the spectrum, our life is merely a blip in comparison to evolutionary or geological processes that can span millions of years. It is very rare that a computational model in systems biology is tasked with addressing time scales exceeding two orders of magnitude at once. There are exceptions, such as the slow accumulation of plaques in our arteries or the very slow clouding of the lenses in our eyes over a lifetime but, by and large, most models address only one or two time scales. This intrinsic restriction of a reduced time horizon lets us get away with approximations, where we would otherwise have to consider very complicated dynamic functions.

Finally, most models in systems biology are constructed to analyze and interpret actual biomedical data, and these data, by their nature, are corrupted by experimental noise and plenty of uncertainties. Furthermore, all features we measure in biology have natural variability, which adds to experimental variations in measurements. As a consequence, biology itself is seldom 100 percent accurate, and it appears that there should be no real need to strive for 100 percent precision in our models either. After all, perfection is possible in math, philosophy, and theology, and possibly in the arts, but seldom in our day-to-day lives. The famous Venus de Milo is considered beautiful not in spite but because of a slight asymmetry in her face. Maybe close don't count in baseball, but close enough is often good enough in real life and especially when we design models in systems biology. In fact, it is the best we can do.

9

Emergence preparedness

Surprises are great for children's birthday parties, but we surely don't like to receive them from our machines. We would find it rather annoying if the remote control changed TV channels on its own, and we would certainly not want to fly in an airplane that surprised us by deciding to start its descent while high over the ocean. Surprises are events that we do not expect to happen. In the more somber language of science, a surprise is quite similar to the notion of an emergence or emerging behavior. Many definitions of emergence have been proposed, but the core concept is that we cannot explain an emergent property of a system by only studying its parts. One does not have to look far to find emergence. Wave patterns on a lake are difficult to explain in terms of individual water molecules. Table salt enhances the taste of many foods in a pleasing manner, but we would not want to taste its constituents, sodium and chlorine. The parts of a clock by themselves do not tell time. Aristotle already wrote in his book *Sophistical Refutations* about this "fallacy of division" and the corresponding "fallacy of composition," which debunks the faulty inference that everything that is true for a part is also true for the whole, and vice versa.

Emergence has had a prominent role for a long time, both in biology and in philosophy. This prominence is not surprising, because no emergence is more stunning than life itself. About 99 percent of the mass of our bodies consists of hydrogen, oxygen, nitrogen, carbon, sulfur, and phosphorus, and roughly 99 percent of all molecules in a typical living cell are water. Yet, if we would buy water and other chemicals, we would still be eons away from a human body. What exactly constitutes the diffe-rence between a functioning cell and its components? What is different just before and after the death of a cell or organism? Is emergence the secret of life? The difference between life and its nonliving parts poses a fundamental, unanswered question that has been keeping philosophers and biologists wondering and pondering for a very long time. Similar persistent puzzles can be found throughout nature, from insect colonies to the relationships between thought, memory, and consciousness and their biological foundation.

At a societal level, emergence is related to the law of unintended consequences. Many governments, agencies, and individuals have tried to manipulate societal issues through targeted interventions that at first appeared to be very reasonable. However, because any society is a complex system, such attempts often failed and sometimes made the situation worse, as Dietrich Dörner vividly describes in his intriguing book, *The Logic of Failure*. Out of well-intended alterations of parts and processes, big problems can arise as unintended emergent properties. Exemplary mishaps include the introduction of rabbits to Australia, which, without natural predators, soon became so successful that they began to endanger some of the native species, and the use of the ivy-like Southeast-Asian kudzu plant as ground cover in the southeastern United States, where the plant is now growing out of control in many locations.

While there is no universal agreement, many philosophers and sci-entists maintain that an emergent property is a feature of a system as a whole, yet not of any of its components. This notion directly reflects what Aristotle famously wrote in his book *The Metaphysics*: "The whole is something besides the parts, ... the whole is greater than the sum of its parts."[1] Somehow, complex spatial or behavioral patterns can arise out of relatively simpler, more fundamental entities, and these novel patterns are intrinsically different, autonomous, and irreducible with respect to their parts. This process of self-organization, arising out of a multiplicity of simple interactions among diverse component parts, has been fascin-ating to thinkers of all times.

The emergence of genuinely new phenomena requires nonlinear fea-tures within the investigated system. In particular, emergence is at odds

1 https://en.wikiquote.org/wiki/Aristotle.

with the important superposition principle of linear systems, according to which the response to two stimuli exactly equals the sum of their responses, as was discussed in Chapter 1. Biological systems very frequently operate in a distinctly different manner. They often exhibit synergism where the response to two simultaneous inputs is stronger than the sum of the inputs when they are applied by themselves. It is also possible that the system exhibits antagonism, where the combined effect is weaker. The consequences of synergisms and antagonisms are difficult to predict in biological systems, but they are very important to investigate. For instance, if a single clinical measurement is slightly too high or too low, a patient may not need treatment. However, if several markers are outside their normal ranges, the situation could become problematic. The side effects of a prescription drug are well characterized before the drug is allowed to enter the market. Yet, if patients take several medicines, their interactions are sometimes surprising and worrisome. Who would have thought that over 40 different prescription drugs can be deadly when taken together with grapefruit juice? We used to think that many outward characteristics of a person were associated with one or a few genes. We now know that very basic features like body size or the disposition for a chronic disease are driven by complicated, and usually unknown, synergisms and antagonisms among dozens, if not hundreds, of genes.

Many scientists have considered the phenomenon of emergence in biology as a strong argument against reductionism. Reductionism divides and subdivides systems successively into subsystems and their ultimate building blocks, hoping to gain an understanding of the system from comprehensive knowledge of its parts. This research strategy runs counter to the study of emergence, which implies that the whole is not only bigger or more than its parts, but also intrinsically different, which proves that knowledge of the parts cannot be sufficient. Biology is more than physics and chemistry, and a cell or organism is more than its molecular inventory. Thus, reductionism is not sufficient, but it is nevertheless crucially important, because we can certainly not understand a system without knowing its parts.

During the first half of the twentieth century, German philosopher Nicolai Hartmann called emergence a "new category" at a higher level of a hierarchical system and attributed it to the cooperation among different types of lower-level elements, which he saw as forming something entirely new. Other philosophers used the terminology of an emergent property that supervenes on the properties of the parts, and thus has an influence in a downward fashion. Because lower-level components alone cannot explain emergent properties, it was postulated that complex chains of dynamic interactions among the components had to be essential.

The contemporary American philosopher Mark Bedau argues against the perennial philosophical puzzle of this type of "strong emergence," although he considers it logically possible. Interested in questions of artificial life, he finds it tantalizing but also quite uncomfortable that features allegedly appear without a solid scientific explanation. In fact, he considers the aspect of "getting something for nothing" in emergence as illegitimate magic. This view reflects famous biochemist J.B.S. Haldane's comment in 1932, that "the doctrine of emergence ... is radically opposed to the spirit of science."[2] One should add that Haldane, ten years later, expressed doubts that one could ever fully answer the question of what life is.

As a partial solution to the dilemma, Bedau proposes the slightly different concept of "weak emergence." According to his proposal, the high-level macro-state of a system has macro-properties that are determined and explained exclusively by interacting micro-states at lower levels and inputs from the environment. Intriguingly, and in contrast to simple causes and effects, Bedau argues that the emergence of macro-properties can only be discovered and explored with simulations. Their necessity is actually seen as a defining feature of emergence and complexity, rather than a sign of technical or human limitations.

Weak emergence of the type Bedau proposes can be explored very nicely with systems of ordinary differential equations (ODEs) and also with rule-based or agent-based models (ABMs). Let's look at the latter first. ABMs generically consist of two features. First, there are agents, which may represent just about any type of entity, from molecules to cells to people. Governing these agents are rules that determine what an entity can possibly do and what happens if two entities encounter each other. For instance, two molecules may form a complex or undergo a chemical reaction that leads to a new type of molecule. The state of the system is successively updated at time points 0, 1, 2, 3, ... but not in-between. It is surprisingly easy to create all kinds of patterns with these models, even if the rules are quite simple. As a simple example, the

2 J.B.S. Haldane (1932) *The Causes of Evolution*, Longmans, Green and Co., London, 113.

rule "turn blue if you are red and turn red if you are blue" immediately leads to a blinking pattern. The rule "no matter what your color, take on the color of the majority of your neighbors" leads to a separation of colors and fewer, larger domains of dominating colors. Rules can generate snowflakes or the stripes on the back of a zebra. They can generate pulsing wave patterns or spontaneously create order out of disorder. A well-studied example of a rule-based simulation is the *Game of Life*, which the British mathematician John Horton Conway invented in the 1970s. It consists of a two-dimensional area resembling a large checker board. Whether a cell is born, stays alive or dies between two time points is determined by a few simple rules. Depending on the choice of the initial number and positions of living cells, the game can generate very complicated, self-organizing patterns, including those that move across the board or form traveling waves.

Self-organization is an essential concept in biology, and examples can be found at every level of biology. Once an amino acid string has been generated by the translation of RNA, this string immediately begins to fold into a protein. Under unfavorable conditions, single slime mold cells use chemical signals to aggregate and form a complex fruiting body that looks like a single organism. A fertilized egg cell develops into a blastula, gastrula, and eventually an embryo. In all these cases, there is no central authority telling the individual cells what to do, or how, or when. Each ant in a colony is an autonomous entity that responds to local signals, such as chemical scents, and leaves traces of its own scent as a stimulus for other ants. Neither the queen, nor any other authority, makes decisions for the colony, and yet, the colony functions very well and is able to respond to never-seen perturbations and adapt to new conditions. Much effort has been devoted to reverse engineering the rules that lead to self-organizing biological processes. In fact, some speculated rules are being used in modern computer algorithms to determine optimal states for complicated systems. An example is an optimization algorithm that tries to mimic the process with which ants find optimal paths toward new food sources.

Relatively simple rules can also create self-similar fractal patterns. Such patterns are independent of the spatial scale at which they are considered. In other words, zooming in on a part of a fractal reveals the same fractal again and again. Branching patterns in arteries, airways in the lung, and the dendrites of neurons are biological examples. Many computer algorithms have been developed for creating sometimes stunningly beautiful fractal patterns.

The emergence of self-similarity also arises in the structure and evolution of biological networks, which, similar to the modern organization of airline routes, often contain a few hubs with many connections and more and more hubs with fewer and fewer connections. The pattern of relatively few, well-connected airport hubs makes it possible to travel from any airport to any other airport with a much smaller number of flights than if all airports were about equal in terms of importance and flight connections. This type of organization in hubs has been termed a small-world network. An example in biology is metabolism, where a few metabolites, such as water and ATP, are involved in hundreds of reactions, whereas the vast majority of metabolites participate in only three or four reactions.

Emergence can be nicely demonstrated with systems of differential equations. In fact, it is often the goal of a dynamic model analysis to identify specific model settings where a property emerges before our eyes. The archetypical example is the Hopf bifurcation. The situation is the following. A dynamical system is residing at its normal steady state, where material is flowing through the system, but none of the variables is changing in quantity, because all fluxes flowing into each pool exactly balance those that are leaving this pool. If one changes a parameter value in the system, the pool sizes may change a little bit, but nothing much else happens. However, if the same parameter value crosses an invisible threshold, everything changes. The system no longer remains at its steady state, but begins to oscillate without ceasing. The required change in the parameter value is infinitely small, but the behavior of the system becomes entirely different. This emergence of a stable oscillation was named in honor of the twentieth-century mathematician Eberhard Hopf, who studied this phenomenon in detail. Another class of well-studied emergences consists of chaotic systems, which are entirely deterministic, yet unpredictable, unless they are analyzed with computer simulations.

Emergent properties may be most striking in actual, complex biological systems, but they can also be found in very small dynamic models, and even single equations. As we will discuss in Chapter 10, a single differential equation is sufficient to model a peculiar, chaotic phenomenon. The equation produces an apparently stable oscillatory pattern but, without any warning, and at an unpredictable time point, the oscillation crashes and starts oscillating around a totally different value, before eventually resuming its original oscillations. The up and down continues in unpredictable patterns for as long as one solves the differential equation. This scenario falls squarely into Bedau's concept of weak emergence, as none of the details or components within the differential equation are hidden, unknown or chaotic, and it takes a computer simulation to reveal its erratic behavior. Larger models may show even more, and more complicated, emergent properties that arise at unpredictable times. However, it is also often the case that large systems possess complex regulatory control structures that buffer against emergent behaviors, which are not always welcome for the normal operation of a cell or organism.

All in all, the situation is very intriguing. First, even simple emergence can confuse our intuition. Second, we can analyze emergence in small, suitably structured systems with rigorous math. And third, we can demonstrate emergence in slightly more complex systems with computer simulations. Nevertheless, none of our tools allows us to understand the emergence of complex functions or patterns, for instance, in the brain. Is it possible that our language of math or of systems biology, let alone our vernacular, is too restrictive or unsophisticated to explain emergence in complex systems? Are our methods of thinking and rational arguing too primitive and limited to explain the emergence of life from molecules, or the emergence of cognition from an assembly of neurons? Alternately, do we have to surrender to the sobering and science defying thought that the emergence of complex phenomena is genuinely something that cannot be explained? Is the problem simply that we cannot measure enough, or is it genuinely deeper? Would it be sufficient to recreate an emergent phenomenon from fundamental building blocks in order to claim that we understand the phenomenon, or does a true explanation require an underlying theory? At present, nobody is truly able to answer these questions.

One fact is clear. Living systems are more than the sum of their constituents, and this "more" can be found in synergisms, threshold effects, and the emergence of unexpected responses. Biology and medicine have made incredible progress, and collectively we know many times more than any other human society in history. Yet, we have something in common with all of our ancestors: we experience, time and again, that life is still full of surprises.

10

Life without chaos?

We all use thermostats, shock absorbers, and various kinds of insurance to buffer our life against strong fluctuations. Yet, if two kids are sick, the dog just ran away, the car has a flat tire, and we are late for work anyway, life is just chaotic. Is it even possible that there is normal life without chaos? If we could ask the Greek theologian and philosopher Hesiod, who lived in about 700 BC, he would have insisted without doubt on the utmost importance of chaos. After all, Chaos was a primeval deity that existed before all the others and gave birth to Love, Silence, and the Night. Allegedly, Chaos even preceded the gods responsible for the underworld and for Earth itself. Chaos was imagined as the gap of total emptiness between heaven and Earth, from which the cosmos was later created. This vision was actually not all that different from the Judeo-Christian version of the beginnings of the earth, which initially was tohu wa bohu: without form and void.

Today, the notion of chaos is rather different. In the vernacular, it refers to disorder or confusion, mayhem and unpredictability. In math, and by extension in the theory of dynamical systems and in systems biology, chaos has been given a much more specific and narrow definition. Namely, it describes the behavior of a system that follows deterministic rules but nonetheless seems to be unpredictable and random. Is chaos possible in biological systems and does it play a significant, positive or negative, role?

To appreciate the definition of modern-day chaos in dynamical systems, one needs to start a little earlier and look at random, stochastic effects, and deterministic processes. A first question, which sounds innocuous but really isn't, is whether there is true randomness. We often associate randomness with a lottery or with betting games, such as flipping a coin, rolling the dice or playing roulette. We call the outcomes of these games random, because we simply cannot predict them, no matter how long we play. At the same time, some aspects are not entirely unpredictable. For instance, if we flip a fair coin 1,000 times, we expect that we should see about 500 heads and 500 tails. If we roll a single die 6,000 times, we expect to find each of the 6 faces roughly 1,000 times. Although the next roll will yield an entirely unpredictable result again, the collective result is very much predictable, at least in a statistical sense.

Is every roll of the die really random? How about a random number generator? A good, intuitive illustration of the situation is the following. Suppose a person is looking out of the window of a high-rise, watching people walking down the sidewalk. Across the street, the observer notices that a flowerpot is right at the edge of a windowsill. All of a sudden it tips, tumbling down toward the sidewalk. If a person is walking along the sidewalk, the observer can predict with some reliability whether this person will be hit by the flowerpot or not. For the observer, the process is mostly deterministic. The location of the falling flowerpot depends on speeds and distances, and possibly on other factors like the wind, and while some stochasticity remains, as the walker may suddenly turn or stop, the scenario is by and large foreseeable for the observer. Not so for the unsuspecting person on the sidewalk, for whom the hit by the flowerpot would appear as really bad luck, that is, entirely unanticipated and random. The scenario demonstrates that randomness is not absolute; rather, the perception of randomness depends on the amount of information the observer and the person on the street have.

This difference also applies to other random processes. Consider a random number generator. It consists of a complicated computer algorithm that spits out numbers, let's say between 0 and 100, with the same probability. A high-quality random number generator hits every number, including quite a few decimals, with almost exactly the same probability and without any recognizable trends. Thus, if the last number was 25.21955, the next number could be 25.21956 or 8.21953 or any other number between 0 and 100. There is seemingly no rhyme or reason for the next number, and we can watch a very long string of these numbers, without ever being able to forecast the next one. However, this being a computer algorithm, somebody has coded it, and if this person knew the present state of the computer and the last numbers, she could forecast the next numbers with certainty. Thus, random numbers are in truth pseudo-random numbers. Finally, let's look at coin flipping or dice rolling. If we could measure hand movements, material properties, and other physical and environmental conditions with incredible precision, it would not seem to be impossible to predict what the next outcome would be.

The moral of the story is that random is very often a matter of information. Whether true randomness exists is actually almost a philosophical question, which can quickly lead to deep questions about determinism, free will, and fundamental theological creeds. Or one may end up in the realm of quantum mechanics and Heisenberg's fundamental uncertainty principle, according to which we cannot simultaneously know the location and momentum of a particle.

In biological systems analysis, these discussions are usually of secondary importance and one just pretends that apparent randomness is truly random, because it is usually impossible to measure all processes that truly drive a phenomenon. Due to this lack of knowledge, the next event appears to be random and is therefore unpredictable. Nevertheless, the collection of many events can be summarily assessed with methods of statistics. For instance, if the random number generator spits out millions of numbers, we are likely to obtain roughly as many numbers below 50 as above 50.

The mathematical field of stochastic processes addresses chains of random events. The word stochastic comes from Greek and refers to a person who is capable of hitting a target, for instance with an arrow. This is quite interesting, because the "stochastikos" is not a gambler, but an expert archer who has learned to control the randomness associated with shooting arrows, at least to some degree.

Two series of events, one from a stochastic process and the other from a chaotic process, can be very similar and may seem to be indistinguishable. However, behind the scenes they are very different, because the next event in a chaotic series is determined by the current state of the system, while the random event is not. In particular, if one records the two series of events at time points 1, 2, 3, …, and makes a plot where the x-axis contains the event at one time point, and the y-axis contains the event at the next time point, the random plot will eventually fill the entire space, while the chaotic system will show a very clear pattern, which is often a fractal. Thus, there is a certain dualism: chaos is entirely deterministic but forever unpredictable in detail, and it is not appropriately characterized with statistical measures such as mean and standard deviation. By contrast, randomness is entirely stochastic, but in the long run "tamed" by statistics. In particular, the collection of many events often forms a well-defined statistical distribution.

It is possible to use a single differential equation to create chaos, but not true randomness. One method of generating chaos with a differential equation is to start with a stable oscillation, called a limit cycle, which keeps on beating without damping or growing, and to poke it with a sine function. For many choices of frequency for the sine function, nothing spectacular happens. For example, if the frequency is very high, the original limit cycle oscillation continues, superimposed with fast wiggles created by the sine function. However, if the frequency is chosen just right, the oscillation becomes very erratic, and the oscillatory pattern never repeats itself; it is always slightly different from all former patterns. This outcome is quite amazing, because absolutely everything in this system is known: the differential equation, its parameter values, the start values for the equation, and the features of the sine function. An intriguing example is the so-called blue-sky catastrophe. The output of this differential equation is a seemingly well-behaved oscillation with a relatively small amplitude and a regular frequency that can go on for a long time. But all of a sudden, without any intervention, like lightning out of a blue sky, the oscillation crashes down to a low level, where it then continues to oscillate regularly. Depending on the numerical settings, it may jump back quickly or only after a long time. A single differential equation is able to model this phenomenon. If we wanted to model a cell, we would need thousands of differential equations, giving us cause to pause and wonder.

The same ability of chaos can be found in difference equations. These describe the state of a system at discrete time points, similar to a digital clock that only shows minutes, but not seconds. In such a discrete system, the state depends exclusively on the previous state. Although entirely deterministic, even a single difference equation may lead to unpredictable, erratic oscillations and chaos.

The behavior of a chaotic dynamical system is unpredictable in detail, but certain statements or predictions can nevertheless be made.

For instance, many erratic oscillations will never venture beyond some lower and upper thresholds, which depend on the numerical details of the system. Typical for chaotic systems is the fact that their specific responses depend very strongly on their initial state. If two chaotic systems start a hair's-width apart, they will nevertheless oscillate out of phase and after a while exhibit totally different patterns. This observation may be seen as another indication that life is chaotic, because how many adults have anything to do with their peers from kindergarten? All the kids started in essentially the same place, but drifted apart, and, while remaining within the confines of the earth, they may have easily ended up in different corners of it. This drifting apart is related to the famous butterfly effect: the system as a whole is so sensitive in its responses that the flapping of a butterfly wing in a distant land affects the details of its future.

Whether randomness and chaos really exist in biological systems is a difficult question that we will never be able to answer, because, mathematically speaking, it may require very long and dense series of measurements to differentiate the two beyond doubt or ambiguity. With hand-waving arguments we will probably conclude that both exist. In fact, nothing much in biology or medicine is strictly regular. In part, the lack of regularity is driven by the constant bombardment of organisms by unpredictable, external events in their environment. But internal events seem to be erratic as well. The most "regular" heartbeat fluctuates somewhat with respect to heart rate and blood pressure even under normal, stress-free conditions. Measurements of brain activity are the most regular in individuals with brain damage. In healthy individuals, the complexity of brain waves increases from quiet sleep to quiet wakefulness to active mental tasks. Chaos has also been claimed at much lower biological levels, such as metabolic pathways and gene expression. Nonetheless, by simply observing systems, we can seldom discern with certainty whether an observed phenomenon in vivo is truly chaotic, stochastic or deterministic.

The difficulty of classifying a phenomenon is independent of the fact that an erratic oscillation in a biological system may have significant consequences. In fact, and surprisingly, these may be good or bad. Let's start with the latter. Cardiologists tell us that most people will ultimately die with (or from) an erratic heartbeat. The normal, healthy pattern consists of 60 to 80 beats per minute, and every phase between two beats exhibits a very distinct shape in an electrocardiogram. If the heart enters fibrillation, different heart regions appear to have their own rhythms, and the overall heartbeat pattern starts to deviate from this normal appearance. While some fibrillations are tolerated, and are not even cause for concern, most electrocardiograms of the last minutes of life show abnormal fibrillation patterns. At the same time, and very interestingly, the heartbeat of a fetus that does not show any chaotic characteristics but is entirely regular is almost always a sure sign of distress and severe problems.

More generally, Lewis Lipsitz[1] and Ary Goldberger[2] proposed that there is a direct association between complexity, chaos, aging, frailty, and disease. They argued that normal physiological functionality is governed by complex interactions of multiple regulatory systems. These systems enable the healthy body to adapt quickly to unpredictable stresses and changes in the environment. Aging and disease may be characterized by an impairment of these processes, which leads to a reduction in the dynamic range of physiological function and therefore a loss of adaptability. In particular, high-frequency components are reduced or disappear, and some processes, which normally show a complicated, possibly chaotic, periodicity, become too regular and lose their complexity. Similarly, the secretion of some hormones becomes weaker with age, and the complexity of some anatomical structures in nerve cells and bones decreases. Lipsitz and Goldberger demonstrated these connections with a number of physiological examples, including heartbeats and brain waves, where losses of complexity occurred. In particular, they proffered that the healthy heart is a chaotic oscillator, and that sudden death may be the transition out of, rather than into, chaos. Going one step further, they suggest that complexity and chaos theory could eventually lead to clinical markers for monitoring senescence and characterizing age- and disease-related reductions in the physiological capacity to adapt.

The final assessment of the role of chaos in living systems is still unclear. Every organism must respond to changes surrounding it, and

1 L.A. Lipsitz (2004) Physiological complexity, aging, and the path to frailty, *Sci. Aging Knowledge Environ.* DOI: 10.1126/sageke.2004.16.pe16.

2 A.L. Goldberger (1990) Is the normal heartbeat chaotic or homeostatic?, *Physiology* 6 (2), 87–91.

these changes are often erratic and possibly chaotic. But even internally, the multitudinous regulatory systems in the body enable complicated oscillations, which may be harmless, neutral, positive or negative. In the context of heartbeats, and maybe of life in general, too much chaos can create real problems, but a little bit of it actually appears to be beneficial.

11

What hath God wrought!

On July 15, 2013, the last telegram was sent by the only remaining telegraph office in the world, the Central Telegraph Office in Janpath, India. The technology had lasted for almost two centuries, since the American artist and inventor Samuel Morse in 1837 first experimented successfully with an electrical telegraph. In 1844, he demonstrated the feasibility of his invention to the world, by sending the message "What hath God wrought" from Washington, DC to his colleague Alfred Vail in Baltimore. By doing so, he ushered in the era of the electrical and electronic information age.

The appeal of sending information across long distances is obvious and, of course, was not new even in Morse's time. The Estonian diplomat Pavel Schilling had demonstrated the feasibility of an electrical telegraph a few years earlier, and French engineers had operated light signaling systems since the late eighteenth century. Long before these inventions, Native Americans had been using smoke signals to warn their peers of impending danger.

Considering the enormous importance of knowing what's going on at a distance, it is not surprising that the transmission of information is an integral part of essentially all biological systems. The specific mechanisms of signal transduction within and between cells vary greatly in their molecular basis, transmission distance, time scale, and a number of other features. For instance, the speed of signal transduction ranges from electrical signals at the order of milliseconds to protein-based signaling cascades responding at the order of minutes and signals involving genomic and physiological responses that occur within tens of minutes, hours or even days.

In many cases, multiple signals are transmitted in parallel. For instance, the demand for a metabolite often triggers feedback signals directly at the metabolic level. These signals modify the activity of enzymes that are responsible for the synthesis of the desired metabolite. They accomplish this typically by slightly altering the physical shape of the enzyme molecules, which occurs within seconds. If this mechanism is insufficient, the availability or activity of other proteins may be changed, which takes minutes or hours. Persistent demands are signaled to the genome, where the expression of appropriate genes is up-regulated and secondarily leads to higher enzyme activities over longer time horizons that may span hours, days or weeks. Intriguingly, metabolic, protein-based and genomic alterations may all be initiated simultaneously.

Due to the diversity of signaling modalities, the great importance of each mechanism, and the fact that signaling is a truly systemic feature, signal transduction processes are of premier interest in systems biology and have been receiving enormous attention in recent years. The possible rewards of understanding signaling in detail are enormous. As just one example, many diseases involve the transmission of wrong signals or the misinterpretation of correct signals. If we could learn to understand these processes in sufficient detail, they could be valuable drug targets. Indeed, this way of thinking about disease is already emerging in the pharmaceutical industry.

Signaling is important at all levels of biology. Even lowly bacteria have means of sensing properties of their environments, such as temperature and pH. The external signals affect a receptor on the outside of the cell and are transduced to the inside, where they are interpreted and processed. As just one example of an important survival strategy, such sensing mechanisms enable bacteria to discern the direction of chemical gradients and respond appropriately. Specifically, the signals are interpreted such that they affect the direction of the rotation of a molecular motor – yes, bacteria have motors! – in the form of a circularly whipping hair or a bundle of flagella that propels the bacterium forward. The mechanism is intriguing: a chemical attractant in the environment lowers the activity of a particular enzyme, which decreases the number of phosphate groups on a specific protein, which in turn changes the direction of the motor rotation. As a result, the bacterium tumbles in response to sensing the gradient of the chemical attractant and begins to swim in a new, more promising direction.

Another fascinating example among many bacteria is the process of quorum sensing, which we are slowly beginning to understand in some detail. Each bacterium emits signaling molecules of a specific type that diffuse into the environment. Other bacteria, usually of the same species, have receptors on their surfaces to which the signaling molecule binds. If there are enough signaling molecules, the binding events collectively lead to changes in the expression of a number of genes, some of which actually code for proteins that synthesize the signaling molecule itself. As a consequence, the recipient bacteria produce and emit the signaling molecules themselves in a process that constitutes a positive feedback loop: the more bacteria are present, the higher the concentration of the signaling molecule in the environment will become. If the concentration exceeds some threshold – that is, if there is a quorum – the population stops acting like a bunch of individuals and begins behaving as one "body," for example by aggregating, swarming, forming spores or a fruiting body, or by assembling into dense microbial layers, called biofilms. Such biofilms are found in most moist pipes or tubes and on all kinds of moist surfaces, including our own teeth. If the participating bacteria are pathogens, their aggregation into biofilms can make them very resistant to drug treatments, which is a severe problem, for instance, in the treatment of cystic fibrosis patients.

If we move up the biological hierarchy a bit, we find another very intriguing signaling system, namely, in yeast cells that we know from baking bread and brewing beer. In the wild, yeast likes to grow on grapes in warm climates, and it is not surprising that the temperature may become uncomfortably high. This heat stress is potentially problematic for the cells, because heat can strongly affect the function, physical shape, and activity state of proteins. It may even lead to the denaturation and abnormal aggregation of proteins, as we can see when frying eggs. The direct alterations can cause secondary problems for other proteins, genes, metabolites, and signal transduction systems.

To avoid such problems, yeast cells have evolved a complicated, multi-pronged heat stress response system that acts simultaneously at several functional levels and consists of fast and long-term responses. The first prong of this system is the activation of heat-shock proteins. One of the better known of these proteins, Hsf1p, is a transcription factor, which is activated through several intermediate mechanisms and responsible for the production of protein chaperones. These chaperones assist in the assembly of macromolecules and the proper folding of other proteins, while preventing the pathological aggregation of proteins.

The second prong of the heat stress response is the relocation of specific transcription factors from the cytosol into the nucleus, where they bind to short DNA sequences, called stress response elements, which lead to the expression of numerous genes.

The third prong is the heat-induced change in the shape of proteins that lead to the synthesis of the sugar trehalose. Under normal conditions, the concentration of trehalose in yeast is very low, but if the cells are heat stressed, they quickly produce huge amounts of trehalose, which protects proteins and intracellular structures from damage. All three responses commence within a few minutes once the heat stress starts.

A fourth prong of the response targets a longer time scale. Within a minute or two of heat stress, changes occur in the activities of proteins that synthesize sphingolipids. Once the concentration of a select few of these lipids increases, they signal to the nucleus that there is the need to up-regulate a spectrum of genes that manage a long-term response to heat. Because transcription and translation take some time, this response is much slower but more forceful than the others. Additional responses consist of the turning on or off of various metabolic and signaling pathways that mobilize energy and suspend activities such as cell division, which might be dispensable under heat stress. Importantly, for each of these mechanisms the cell has a corresponding "brake," which stops the response in case the heat returns to normal. Collectively, these response processes form a very complex, highly regulated, and well-timed network of mechanisms that are kept under control with a host of checks and balances.

Cells in higher organisms are involved in many different signal transduction processes, which may be triggered by chemical, physical, and mechanical impulses or stresses. Neighboring cells regularly communicate with each other through channels connecting their cytosols and permitting the movement of messenger compounds such as calcium. This communication occurs throughout the body. An important organ where this type of signaling happens every minute of every day is the heart, where an electrical-chemical signal moves as a wave across the heart muscle and causes well-coordinated contractions of the different heart chambers.

A very common signal transduction mechanism within higher cells is a signaling cascade. Typical cascades have three levels, and it has been shown with mathematical means that three is functionally the charm; two levels are not as efficient, and four levels are not needed. Each layer consists of a protein that can either be phosphorylated (that is, has a phosphate group attached in a well-defined location) or not. If the protein becomes phosphorylated, it is usually in its active state and signals that the protein at the next layer of the cascade should be phosphorylated.

The process usually starts with a chemical compound outside the cell that binds to a receptor. The cell has numerous such receptors on the outer layer of its membrane, each binding to specific ligands, that is, particular molecules in the cell's immediate environment. The ligands may come from the cell itself, from close-by cells, or from cells at distant places within the body. The typical example of the latter is a hormone, which is produced by specific cells, glands or organs and sent through the bloodstream to select target cells. Binding of sufficiently many signaling molecules to the cell surface receptors triggers a signaling event. Each receptor is a protein that reaches through the cell membrane and triggers the phosphorylation of the protein at the top level of the cascade. If it is strong enough, the signal very quickly moves from layer to layer, phosphorylating the next protein in line. If the signal disappears, the protein eventually becomes inactive again.

Each level of the cascade has the potential to amplify the signal, so that the ultimate signal is quite strong, even for an initial signal of small magnitude. To make the process more effective, the cascades are often organized along protein scaffolds, which act as assembly lines and speed up the signal transmission by bringing relevant molecules of a reaction close to each other. The signal at the final layer triggers a cellular response. As a typical example, it may cause a transcription factor protein to move from the cytosol to the nucleus, where it binds to the start site of a gene and thereby triggers its expression. The gene, in turn, codes for a protein that is beneficial for the response to the original signal the cell received. In many cancers, a defect in this cascade system can lead to faulty signal transduction and, subsequently, to uncontrolled growth. On the positive side, several drugs have been developed that interact with intracellular signaling cascades and have a beneficial therapeutic effect that outweighs the side effects. Signal cascades can respond to a wide variety of external triggers, such as cytokine proteins that are a sign of inflammation, growth factors, hormones, mechanical forces, and various physiological stressors.

While the principles of the signaling cascade make intuitive sense, the details of the signal transduction system are actually quite complicated. First, the signal is usually very weak, which has several immediate consequences. On the one hand, weak signals may be missed, even if they are important, and, on the other hand, they are easily confused with random noise in the form of chemically similar molecules that happen to come by the receptors. The system must be able to detect these subtle differences. Once the signal is identified as true, it must be amplified inside the cell, which in typical cases is accomplished through the cascade structure. This design is apparently very efficient, because it seems to be present in all higher cells and even in a few bacteria.

Of course, real life is much more complicated than a three-layered cascade, and systems biologists have been intrigued by how the simple base structure can be used for more complex structures that ultimately have the capacity to respond appropriately to a variety of different stimuli. An important variation on the theme is the existence of several parallel cascades, which are composed of different receptors and proteins. These parallel cascades have their own prime responsibilities, but they do not necessarily operate in isolation and instead communicate with each other via cross-talk. Thus, a receptor primarily triggers cascade A, but one of the downstream proteins of this cascade may secondarily activate cascades B and C, and the collective output of all three cascades is interpreted by the cell within its current physiological state and context. This cross-talk is very interesting, because it seems to increase the reliability and fidelity of signal transduction and can furthermore lead to input–output relationships that a single cascade cannot generate. An example is a cellular response that responds exclusively to signals within a certain range; if the signal is either too weak or too strong, nothing happens. Another example is the response to either signal X or signal Y, but not to both at the same time. Known as "exclusive or," this is a complicated task. Detailed analyses have suggested that normal and cancer cells often differ very little in their signaling components, but exhibit clear differences in the way these components communicate with each other through cross-talk. One should also notice that the functioning of signaling cascades can be modulated by relatively simple molecules, such as calcium, nitric oxide, cyclic AMP, and special lipids like the sphingolipid ceramide.

At the physiological level, the most obvious transmission of information from the environment is achieved with our senses, and although these respond to very different physical and chemical stimuli, the mode of information transmission and interpretation is ultimately a combination of electrochemical signals. A typical example of such "mixed signals" is our response to food or hunger. When we have not eaten for a while, the cells in our stomach lining, the pancreas, and the intestines produce the hunger-stimulating hormone ghrelin. Ghrelin circulates through the body and can bind to receptors in numerous tissues. During hunger, the main target is a specific region of the hypothalamus, whose cells control food intake. These cells receive and interpret the ghrelin signal and interact with the brain's reward system, which is controlled by neurotransmitters like dopamine. As a consequence, the brain senses hunger, and eating comfort foods feels good. Once the body senses satiety, another hormone, leptin, acts on the same region of the hypothalamus and signals the body to suppress appetite and stop eating. Like all systems, the hunger system can get out of balance, and mice treated with ghrelin become very obese. Also, it was shown that people with a genetic disposition for lower numbers of dopamine receptors find food more attractive, motivating them to eat more in order to receive a sufficient dopamine "reward."

The receptors involved in the hunger-response process and many other signaling processes belong to the enormously important class of so-called G-protein-coupled receptors, for whose discovery Alfred Gilman and Martin Rodbell received the 1994 Nobel Prize in Physiology or Medicine. The receptors reach from the outer surface of the cell to the inside where they form intricate complexes with G-proteins. When a ligand binds to the receptor, the internal conformation of the complex of G-proteins changes, and this change triggers a signaling cascade. After some recovery period, the G-protein complex recovers and is ready for the transduction of a new signal. The functionality of very many drugs is associated with G-protein-coupled receptors, and the complexity and dynamics of events surrounding G-proteins and their receptors are prime targets of systems biology.

Hormones are produced and received within the same organism. Their close cousins, pheromones, are emitted to attract or inform other members of the species by affecting specific neuronal circuits in their brains. These long-distance signals are of enormous importance, especially within the world of insects, and are used to attract sexual partners, warn the population of danger, call for an aggregation of individuals or make other members of the clan aware of food sources. In a way, the substances emitted by bacteria during quorum sensing may be considered pheromones. And so the circle closes.

We live in an era that has been coined the information age. Indeed, much of the information that is available anywhere in the world is at our fingertips almost all the time. Biological systems have addressed the need for sensing the environment and for sharing pertinent information throughout evolution, from the simplest microbes to the most developed mammals and plants. And whether it is a hunger signal or the response of a yeast cell to heat, the variety and complexity of mechanisms for sending, sensing, and interpreting signals are astounding. At the same time, we are beginning to recognize that there is often a remarkable conservation of fundamental principles and signaling modalities across species. Fast and slow signals, responses within seconds or extended over weeks, often occur simultaneously and in parallel. Evolution has forced these processes to be well coordinated, and this coordination requires sophisticated control structures and numerous checks and balances, which systems biology is presently only understanding in unduly coarse granularity. Improving this understanding is a lofty but ultimately very rewarding goal, because signal transduction is so central to life that the option of manipulating it in a targeted, predictable manner will have unlimited potential.

12

Tell me with whom you go and I'll tell you who you are

It is an interesting exercise to surf the blogosphere and other, similarly reliable sources on the Internet in search of the origin of this saying. It's a Spanish quote. It's Russian, Arabic, Mexican. One blogger gives credit to his mother. Several websites assure us that the famous Greek bard Euripides proclaimed it in his tragedy *Phoenissae* (c. 410 BC) about a group of Phoenician women caught up in a fight among Oedipus' sons for dominion over Thebes. As Euripides said, "every man is like the company he is wont to keep." Some bloggers attribute the quote to Confucius' *Words of Wisdom*. Others go even further back, crediting the saying to the Assyrians. It seems that we may never know the true origin. But whatever its long history, the widespread usage of this morsel of wisdom points to the universality of the deep human appreciation of companionship and the unique character that bonds a group of friends or associates. Whether it is membership of a club, guilt by association, Benjamin Franklin's pithy comment during the signing of the Declaration of Independence that "we must all hang together, or assuredly we shall all hang separately," or the astounding popularity of social media: the company one keeps defines one's being and identity.

The observation that connectivity contains genuine information has not escaped the attention of systems biologists, who are utilizing this relationship within the realm of biological networks. We are well aware that nothing in biology occurs in a vacuum and that every component is connected to many others. The investigative challenge derives from the fact that biological networks and their connections are often only vaguely known. Because it is the task of scientists to discover the unknown, the rationale is the following: if the connectivity within a human group contains information about its members, the same could be true for biological networks. If so, it should be possible to extract novel information regarding unknown biological components from the analysis of better known components with which they associate.

The best-characterized associations and networks among biomolecules are formed by metabolites, due to two facts. First, biochemists had been working on metabolites long before DNA was identified and proteins could be manipulated with any efficiency. In fact, one of the important roots of systems biology is the work of nineteenth- and twentieth-century biochemists who formulated mathematical models for chemical reactions in metabolism. Second, in contrast to proteins, which can interact in all kinds of ways, for instance through attaching or docking to each other, changing the activity of other proteins through the addition or removal of molecular side groups, or even cannibalizing their peers, each conversion of one metabolite into another is strictly constrained by well-known laws of thermodynamics and chemistry. In particular, mass does not appear or disappear in a cell, which helps enormously with bookkeeping and tracking the whereabouts of metabolites. Yes, glucose may be used to synthesize an amino acid, but the conversion simply cannot be accomplished in a single step. Rather, we understand quite well how the six-carbon glucose molecule is cleaved through glycolysis into two molecules containing three carbons each, and how these smaller molecules are further modified and reassembled step by step into the various compounds a cell needs. Usually aided by enzymes, each reaction adds or removes molecular structures in a well-understood manner, and the possible fates of all metabolites and reactions are firmly determined by chemistry and physics. For historical reasons, metabolic reaction networks have been organized into pathways, such as glycolysis, the citric acid cycle, and the purine and pyrimidine pathways, but these pathways are in reality only major transformation routes, from which other reactions branch off. They serve as a coarse interstate system, which is much more densely connected through secondary roads. If a new metabolite is discovered, the lists of possible sources for production and means of degradation are short, because its chemical structure defines its potential neighbors within the metabolic network.

One intriguing question regarding molecular networks is whether they have genuine structural features that are not unique to metabolites, proteins or genes. Systems biologists address this kind of question with methods of graph theory, a field of mathematics that investigates structures consisting of nodes and edges. Examples of graphs are the electric power grid, networks of telephone lines, and distribution systems for commodities. In biology, the nodes may be proteins, genes, pools of metabolites, cellular structures or other sets of molecules or larger entities, while the edges represent transport processes, structural conversions, signals, physical interactions or various other types of functional connections between them.

An obvious feature of a network is how many edges enter or leave a particular node. In social media, the number of friends or followers is enormously important to many folks and without doubt the most prominent metric of bragging rights. This situation is not all that different in systems biology, where possessing many edges is interpreted as a measure of a node's importance. Pyruvate is an important metabolite because it can be produced in various ways and serves as the starting point for many pathways and compounds.

A related generic question is the following. Is each node associated roughly with the same average number of edges? Are edges randomly distributed among the nodes? Do some nodes have many edges and others none at all? The answer for many biological networks is that the edges are neither similarly nor randomly distributed. Instead, most nodes are rather sparsely connected, whereas a few nodes have very many edges and thereby form clusters within a less connected landscape. Water and cofactors like ATP and NADH participate in very many metabolic reactions, whereas rare metabolites are only connected to a few other metabolites. We find a similar arrangement in our system of airport hubs and flight routes. This so-called small-world network organization, which consists of a few well-connected hubs and many less-connected nodes, has definite advantages. Of special importance, it is optimized in a sense that one can traverse the network from anywhere to anywhere else in a surprisingly small number of steps. Thus, an input metabolite can usually be converted into important biochemical compounds with only a few reactions. Small-world networks are also very tolerant to random perturbations, because these will hit sparsely connected nodes with a much higher probability than the few hubs. On the flip-side, knocking out a hub has severe consequences.

Metabolic pathways are very special because they are stringently constrained. By contrast, other biological components may associate in a variety of ways. Not surprisingly, interacting components are frequently found in the same location at the same time. At the genome level, this co-localization can be seen very clearly in bacteria, where genes are arranged next to each other in so-called operons if their products have related functions. An operon is organized such that it begins with a control region, followed by genes coding for the tasks that are to be coordinated. A nice example is the *Trp* operon, which contains genes that code for components of the enzyme that is responsible for the synthesis of the amino acid tryptophan. The control region, which precedes these genes, contains several specific stretches of DNA that serve as regulatory domains. One domain contains a binding site for a promoter protein. If the control region did not contain other elements, then the presence of the promoter would always lead to gene transcription and ultimately keep producing tryptophan. However, the control region also contains a repressor gene, which codes for a protein that blocks transcription of the other genes in the operon. This repressor protein is only active if tryptophan is already available in sufficient quantities. As a consequence, this little system is a self-controlling production machine that generates tryptophan exclusively at times when it is in demand.

Operons are usually rather small, because they are responsible for just one specific task. At the next higher level of organization, they are connected through regulons, which are groups of genes or operons that are under the control of the same regulatory protein. In contrast to an operon, the genes of a regulon are not co-localized but may be distributed throughout the chromosome, and their association network is therefore based not on physical vicinity but on function.

What regulates genes that are not organized in operons or regulons? Systems biology is working intensely on answers to this important and complicated question. One key feature of this control is another network. It consists of proteins that serve as transcription factors. These factors bind to specific control sites on genes and initiate their transcription. The target genes are typically not co-localized, but that is not a problem because the transcription factors are mobile. In higher cells, they can be moved from the cytosol into the nucleus and back, and this translocation is a very important control tool.

Transcription factor networks are anything but democratic. Rather, they seem to be strictly hierarchical and organized in several levels of control, so that a master transcription factor initiates the production of several transcription factors below it, and these factors control other factors even further below. Many genes respond to several transcription factors, which are under the control of different higher-level transcription factors. As a consequence, the hierarchical organization of overlapping transcription factor networks results in very many finely coordinated combinations of genes that respond to stimuli and stresses with altered expression.

Overlapping with the control function of transcription factor networks are networks of small RNAs. These short nucleotide strings are coded in DNA and have the ability to silence the expression of target genes. Although small RNAs are quite a recent discovery, from research on petunias, one might add, we already know of the existence of thousands of them. Because many small RNAs have the capacity to silence several genes, the functional network of their interactions with target genes is enormous in size and complexity.

While systems biology is beginning to understand the hierarchies and specific roles of transcription factors and small RNAs, many questions are still to be answered. Two methods are currently being used. As often in biology, one works from the bottom up and the other from the top down. The former method attempts to identify all genes that have binding sites for specific transcription factors or small RNAs. The latter analyzes the results of genome-wide experiments that measure gene expression in response to many different stimuli and try to identify groups of genes that are frequently co-expressed and might therefore be complementary in their functions. It is the paradigm of a guilt-by-association approach!

In comparison to metabolites and genes, the associations among proteins may take very different forms, because proteins play so many diverse roles. In addition to serving as enzymes, proteins modulate many processes, transduce signals, assemble into scaffolds along which various cellular processes occur, facilitate material transport, and ensure the structural integrity of cells and organisms. To discover new associations between proteins, one often begins by assuming that it is probably not a coincidence if two components of a cell are frequently found close together. This co-localization can be assessed very precisely with fluorescence microscopy. Two or more types of target protein are labeled with different fluorescence markers, and these light up selectively when stimulated with specific wavelengths of light. Close vicinity of the different fluorescing labels indicates co-localization of the target proteins, and some modern imaging methods can measure whether two protein species actually interact. It is possible that the protein species indeed interact functionally or that they simply co-occur, but have nothing much to do with each other. Recent variants of fluorescent methods can sometimes distinguish between these alternatives.

While co-localization experiments with fluorescence microscopy yield valuable and very detailed insights, they are arduous and difficult to scale up to larger systems. The question thus arises whether one might be able to use high-throughput methods, which have been so successful in the field of –omics. This question is being addressed in a new sub-field of systems biology that, unsurprisingly, immediately received its own –omics moniker: interactomics. Attesting to the great importance of protein–protein interactions, many techniques have been developed to study interactomes in a large-scale manner. Unfortunately, the experiments cannot yet be performed inside living cells, resulting in a high rate of apparent interactions that in reality never occur; for instance, because two proteins are not in the same location within a cell, or the cell synthesizes them at different times. An additional challenge that is being tackled with the help of computer scientists is the analysis and interpretation of results. Because a cell contains thousands of proteins, the theoretically possible number of interactions is enormous, and even if only fractions of all potential interactions actually materialize, there are still very many of them, and they are difficult to understand and visualize. As a side note, similar techniques are utilized to understand the interactome of all neurons in the brain. Not to be outdone by protein chemists, neuroscientists coined the term connectome for this specific purpose of determining which neuron is talking to other neurons close by or at a distance. Their hope is that knowledge of the connectome might shed light on the complex functionality of the brain.

A special class of proteins serves the transduction of signals within cells. A cell receives an external signal in the form of a messenger molecule that binds to a receptor protein on the outside of the cell membrane. The binding causes the receptor to change its shape, and this shape change triggers a signaling cascade, which amplifies the signal and ultimately leads to a cellular response, if the external signal was strong enough. Most signaling cascades consist of proteins, and it was initially thought that each cascade had its individual role and functionality. Alas, the inner workings of life are complicated, and it turns out that there is a lot of cross-talk among the cascades. In other words, a signal triggers one primary cascade, but other cascades respond as well. As a consequence, the cascades form an intricate causal network that can distinguish very many different types of input signal and ensures appropriate responses.

The cross-talk within this network begs the questions of exactly which signaling cascades are alerted in a specific situation, and whether there are primary signal transduction routes within the network. Systems biology addresses this question with sophisticated statistical methods that attempt to infer the connectivity of a network from experimental data. While the theory is complicated, the concept of these methods is not difficult to understand. It is again built upon guilt by association. Ultramodern techniques of molecular biology render it possible to tell which signaling proteins are activated in a single cell. By contrast, the connections between the different signaling molecules are not known and cannot be measured. The typical experimental investigation consists of exposing the cells to many inhibitors that act upon different signal transmission steps. By measuring in each case which signaling proteins are affected, these experiments establish which proteins are downstream of each inhibited step. The raw result is a long list of associations between inhibitors and signaling proteins, and the statistical analysis consists of inferring from these data which proteins are most likely to be connected to each other in a direct and causal fashion.

The computational inference of connections within signaling networks is merely the tip of the iceberg. It is an important step toward understanding biological networks and systems, but there is much more to do for future systems biologists. The first critical feature is that biological systems are always regulated. This fact implies that characterizing just the nodes and edges of a network is insufficient; one also needs to characterize what modulates the strengths of their connections, for instance through feedback inhibition. Indeed, such regulatory effects can easily result in dynamic system responses that are impossible to foresee from the network connectivity itself. Moreover, the properties of regulatory systems may change over longer time horizons. Concentrations and magnitudes of fluxes through the systems may adapt to new environmental conditions, and connections may be formed or broken. These time-dependent features make assessments of functional associations much more difficult than for static networks.

Investigating the dynamics of a highly connected and regulated biological system may be compared to the attempt to understand the traffic patterns and bottlenecks in a city during rush hour. A map provides the static network of streets, and repeated traffic reports offer a glimpse into typical flow patterns and problems that occur on a regular basis. Studying the details of changes in traffic patterns after accidents or blockages provides some idea of how the "system," that is, the drivers of the city, respond, and which other routes are likely to be affected secondarily. In order to understand adaptation of the system and predict slowly evolving trends in bottlenecks that might be caused by a growing population or the construction of a new mall, a short series of snapshots is insufficient and only long-term analyses of traffic patterns have the capacity to provide reliable explanations.

Associations, whether assessed per co-localization, co-expression or similarity in responses, do not directly imply functionality or causality. Nonetheless, finding out who is in cahoots with whom often provides the first valuable hints regarding the inner workings of a biological system.

13

Time for a change!

More than 5,000 performances and translations into 17 languages leave little doubt that the off-Broadway musical *I Love You, You're Perfect, Now Change* hit a nerve with the theater community who vicariously enjoyed the trials and tribulations of their fellow humans in the ever-changing dating and mating turmoil of the late twentieth century. Life may be good, but there is this nagging feeling that something better could be just around the corner. Constancy is comfortable and calms the soul; change is exciting. Indeed, we know that nothing really stands still and that all is in flux, as Heraclitus of Ephesus proclaimed 2,500 years ago.

Except for the "Love You" part, it seems that nature has the same attitude toward its subjects and forces them to change without end. All species are in some way optimized and perfect, because they are clearly more successful than their competitors and survived for generation after generation where others did not. But this "optimal" is merely a current "optimal." It resulted from unceasing change since life began and will most certainly transform into a new "optimal" in the future. Nature never stops exploring new options, and the only constant throughout the eons seems to have been change.

We all grow and age, and we really do not notice significant changes from one generation to the next, except, of course, that we are much more knowledgeable and sophisticated than all generations before us and most certainly incomparably wiser than our younger peers. But in spite of all evolutionary events, most of us have two eyes, two kidneys, and all those features that make us humans, and these just don't change, at least not fundamentally. Yes, we know that our roots go back to Neanderthals, Denisovans, *Homo habilis*, and early African hominids like Lucy, but those distant relatives disappeared a very long time ago. Evolution is a topic to be pondered among paleontologists and does not seem to have an urgent role in daily life.

It is actually not all that difficult to observe evolution at work. One only needs to start with a single bacterium and let it divide and multiply. All future cells should be the same, as there is no new source of genes. However, if we check merely a few weeks later, it is very likely that the genetic makeup of the population has become diverse. We used to talk about "wild type strains," but among many of today's microbiologists the term evokes the quaintness of a Norman Rockwell painting. The unrelenting change of wild type strains actually creates challenging problems for scientists who always like "controls" to which variations are to be compared. If these controls themselves change, such comparisons become problematic. Change happens in nature continuously and at all scales and levels, from never-ceasing physical and chemical processes at the molecular level to the replacement of old cells with young ones; from the growth of organisms to changes in populations; from the emergence of new strains to the extinction of species.

Change poses a very interesting challenge for biologists, because essentially all experiments yield results that are merely snapshots of certain aspects of a phenomenon. Such snapshots can be very complicated – just look at the result of a high-throughput –omics experiment with thousands of data points – but they are still frozen in time. Even a high-resolution movie of a cell does not explain what molecular mechanisms make things happen internally. Yet, it is our job as scientists to find out what invisible mechanisms are at work and how they lead to the snapshot pictures we can obtain.

The need to bridge the gap between snapshots and deep insights into cellular strategies requires cognitive processes of inference and integration that are different from wet-lab experimentation, and directly suggests the utilization of conceptual or computational models. All scientists develop conceptual models of what they study, and biologists are no different. Going beyond conceptual models, systems biology suggests the formulation of complicated phenomena as computational models. If designed appropriately, these dynamic models of complex systems have the genuine capacity to weave snapshots into coherent storylines that have the potential to explain how nature solves complicated tasks.

Maybe interesting at first, although all too well known to veteran modelers, one often learns more from a "failed" modeling attempt than from a successful one. Sure, if the model does what one expects and matches the data, well, the modeler is happy. However, if the model is technically correct, but its output differs substantially from the observations, it is almost certain that there are significant gaps in our understanding of how the various components of the modeled system are connected to each other and what their specific roles are. Usually some head-scratching ensues, along with a lot of revisiting of assumptions and model settings. Diagnosing the specifics of the discrepancies often actually leads to clues suggesting how to change the model in the right direction. Are important components lacking? Are processes represented incorrectly? Are some of the parameter values wrong? Is regulation missing? Even assuming that no technical errors are the culprits, a model of moderate size is seldom biologically correct without a lot of revisions, and the cycling between testing the model against data and refining its features is the bread and butter of computational systems biology, especially when it comes to modeling biological phenomena that change with time.

It is noteworthy, although not really surprising, that many changes in nature are very tightly regulated by amazingly complex control programs, which guide these changes in an essentially deterministic manner. In stark contrast, the scientific community is convinced that it is random mutations and rare stochastic events, sometimes called black swans, that are the dominant forces of evolution, and thus of long-term change. As we will discuss later, this reliance on randomness is intriguing, because most mutations, by a long shot, are not to the advantage of the organism. As humans with a single life we would not gamble so ruthlessly, but nature has the luxury of very many individual life stories, and if the miniscule chance of improving a species is to be paid for with a large number of failures, then that seems to be a risk worth taking. As Thomas Edison famously said, "I have not failed. I've just found 10,000 ways that won't work." It may not even be clear-cut how to define "failure" in a natural system. An intriguing example of this situation is sickle cell anemia, a terrible disease for the patient, but a protecting mechanism against malaria that enhances the chances of survival for the population.

Let us start discussing this dichotomy of determinism and stochasticity with the daily routines of life, where growth and change happen according to a plan and where variations are not really welcome. Most growth of significance is achieved by an increase in the number of cells, whether in a free-living cell population or a tissue, rather than the growth of a cell in volume. The fundamental process driving this increase is the cell cycle, in which a typical cell executes its physiological role for a while, but at some point stops, doubles its genetic material, and ultimately divides into two daughter cells. The cell cycle in bacteria may be completed in 20 minutes, muscle cells in the human rib cage can live for 15 years, and some cells in the brain are apparently never replaced. The two daughter cells emerging from a cell division grow in volume until they reach the size of the mother cell and eventually become mother cells themselves. Repeated divisions lead to a rapid increase in the number of cells in a population: without losses, just 20 divisions lead to over a million cells.

Even in very simple organisms, hundreds of genes are involved in the control of the cell cycle. This fundamental process of life is so complicated because all types of cellular component within the mother cell have to be doubled and separated, so that each daughter cell is complete and receives its fair share. Many parts of the inheritance can be divvied up without overly great precision: if one daughter cell in the liver receives 820 mitochondria from their mother and the other one only 800, they will both do just fine and eventually adjust these numbers. However, there is one portion of the cell that does not permit any laxness, namely, the genome. Every gene is important, and for a daughter cell to receive either fewer genes or more genes is problematic in both instances. If genes are missing, some functionality is missing, and the cell incurs a high risk of malfunctioning or dying. At the same time, we know that a single additional chromosome can lead to Down's syndrome and that some cancers are associated with gene duplications or insertions of DNA in wrong locations. We also know that mutations, even if they only cause slight changes in the building blocks of DNA, are incomparably more often disadvantageous than beneficial. So, the cell has to make sure that every piece of DNA is copied as correctly as possible. But that seems to be insufficient: in addition to ensuring almost error-free copying, nature has developed various DNA repair mechanisms that drastically reduce wrong gene sequences.

To put this task in perspective, while ignoring repair mechanisms, consider a symphony orchestra performing a Beethoven symphony. Coarse estimates suggest that the conductor's score contains about 60,000 notes. Because all second violins are supposed to play the same notes, and the same is true for first and second flutes, trumpets, and double basses, the orchestra plays a total of roughly 300,000 notes. If any one of the musicians plays a single wrong note during the concert, the "error rate" for the orchestra is 1 in 300,000, give or take a few thousand. Now compare this number to our human genome, which consists of about 3 billion nucleotides that are the base units of DNA and need to be copied faithfully before each cell division. The error rate here varies a bit among different tissues, but has been estimated to be at the order of one in a billion. To achieve this level of accuracy, the orchestra would be allowed a total of one wrong note in about 3,000 performances.

Although cell cycles are very complicated, dedicated modeling efforts by the group of John Tyson and others succeeded in capturing and integrating much pertinent experimental information in the form of mathematical models. These models have given us quite a good understanding of the molecular basis of the cell cycle and, in particular, of the involved genes and proteins, their roles, and finely timed interactions.

While many cell divisions lead to very similar daughter cells, a large class of divisions is programmed to be asymmetric, thereby resulting in two substantially different daughter cells. Capturing and understanding the interplay between the multiple factors and mechanisms governing and regulating this process are true examples of what systems biology is all about. A crucially important case of such an asymmetry is the process of embryonic differentiation, which is a key driver of organismal development, because a single fertilized human egg cell is the sole source for ultimately about 100 trillion cells of roughly 400 different types in the body.

In many differentiation processes, an asymmetric cell division leads to one daughter cell that is very similar to the mother cell and to one "new" type of cell. In somewhat simplified terminology, the mother is a stem cell that divides into two daughter cells. One of these is again a stem cell like the mother cell, while the other daughter is a specialized, differentiated cell. The differentiated daughter cell can have very different properties than her mother and sister in terms of size, shape, responsiveness to signals, and many molecular features. While the fertilized egg cell is the ultimate stem cell, from which all cells in the body derive, some adult stem cells can also lead to different cell types; an example is the lineage of white blood cells.

Of note is that even in asymmetric divisions, both daughters and their mother have exactly the same genome. So, where do the stark differences between the daughter cells come from? The current thinking is that the new features of the differentiated daughter cell are the result of an unequal distribution of regulatory molecules in the mother cell during the separation into the two daughter cells, and that these regulatory molecules switch on different blocks of genes. It also seems that the daughters might differ in "epigenetic" attachments of molecules to the DNA that alter gene expression. Many details of the complex mechanisms controlling gene expression during differentiation and of the important role of stem cell-specific epigenetic modifications have been slowly emerging, but we can certainly not claim to comprehend this miraculous switch completely. It seems that hierarchies of transcription factors, gene promoters and enhancers, repressor proteins, as well as regulatory RNAs are involved in the differentiation process and coordinate the expression or suppression of entire arrays of genes. Moreover, the DNA of higher organisms is wrapped around certain proteins called histones, and these can be modified during differentiation to control access to transcription factors, thereby resulting in changes in gene expression.

Differentiation requires enormously precise coordination, which is mediated through targeted alterations in the expression of appropriate genes. However, it is also affected by the physical and chemical surroundings of the differentiating cell. It is actually possible in the laboratory to differentiate a stem cell artificially into different cell types by using different concoctions of chemicals in the medium. This observation leaves no doubt that mechanisms of signal transduction between cells, or from the environment to the genome, are key processes. Many of these signals are transmitted by small molecules that are often called growth factors or morphogens – "shape makers" – which bind to receptors on the surface of target cell types. Binding alters the shapes of the receptors, a process that transmits the signal to internal signaling cascades, which in the end causes the relocation or activation of transcription factors in the nucleus, where they turn on the targeted genes. Differentiation can even be affected by the physical surroundings of a cell, such as the stiffness or coarseness of a surface on which the cell grows.

Differentiation processes are not only crucial during embryonic development and wound healing, but happen millions of times every day in tissues that incur heavy wear and tear, including our skin, the liver, the lining of the gut, and many others. As a specific example, consider the unceasing production of white blood cells in the bone marrow, which requires several differentiation steps. A cluster analysis of about 6,000 human genes identified about 570 genes with significantly altered expression during this multistep process, and these fell into 12 groups of genes that were active during different phases of the cell cycle. In one of these groups, the expression of 28 genes was slowly turned on at exactly the same time when the immature white blood cells turned into macrophages and lost their ability to proliferate. The specific roles of some of these genes could be explained with our current knowledge, but others were puzzling and pointed to significant gaps in our understanding of the system.

In some cases, differentiation even causes one of the daughter cells to commit suicide. While at first maybe surprising, this process of apoptosis or "programmed cell death" is of great importance. For instance, during the formation of fingers in an embryo, cells must die to allow the fingers to separate.

Future studies of the very many aspects of differentiation will involve traditional as well as –omics experiments, which will reveal ever more details. Methods of bioinformatics and machine learning will organize the experimental findings and show which data are most informative. Integrating the growing body of information into dynamic models may eventually give us a more comprehensive understanding of how it is possible that a single fertilized egg cell can turn into 400 different, healthy cell types.

Because embryonic stem cells can differentiate into every cell type in the body, intense recent research has been trying to reverse the process, that is, to "de-differentiate" adult cells, such as fat cells, back to stem cells that have more or less the same range of possibilities as embryonic stem cells. The rationale is that these "induced pluripotent stem cells," or iPCSs, could then be differentiated into person-specific tissues that could replace those damaged by aging or disease. For example, it would be of enormous significance – and profit – to replace diabetic pancreatic tissue that had lost its capacity to control sugar levels in blood appropriately. Some successes in this direction have already been spectacular, but much more can be expected in the next decades. The process of de-differentiation would also offer a viable alternative to the use of embryonic stem cells, which is frequently considered unethical.

It is possible that cells dramatically change even without a cell division. They may experience changes in shape, size, appearance or functionality in response to unfavorable surroundings or clues from the environment. For instance, many microorganisms can form spores that have a tough "shell" that protects them from a large number of threats. Some viruses can either infect a cell, multiply many times, and then burst out of the cell to infect other cells, or "hide" for a long time by inserting themselves into the DNA of the host, where they are copied every time the cell divides, only to become active, multiply and burst the host cell at a later time.

A particularly intriguing example of a change without cell division is the so-called epithelial–mesenchymal transition, which unsurprisingly is usually referred to as EMT. During this process, an epithelial cell, which is part of the lining of a blood vessel or a cavity in the body, gradually loses some of its characteristic features, such as its ability to stick to other cells. It becomes mobile and moreover regains some properties of a stem cell that can differentiate into other cell types. EMT is crucial during embryonic development and wound healing, but it is unfortunately also a key process in the spreading of tumors throughout the body via the blood stream. What triggers EMT in cancer is unclear, but EMT involves several signaling pathways and coordinated changes in the expression of probably several thousand genes. EMT is reversed in a process called MET, which in the case of cancer can lead to the formation of a metastasis.

The duplication of DNA during cell division is extraordinarily accurate, as we discussed before. In addition, cells possess proofreading mechanisms that trigger the destruction of wrongly copied DNA. Nonetheless, mistakes happen and sometimes go unnoticed. In most cases, copying errors are actually not all that critical. They usually occur in cells of the body that are easily replaced, and no real harm is done, although the error can possibly lead to disease. For the individual, this outcome can be devastating, but for the population, errors are much more important elsewhere, namely, in the generation of germ cells. If an egg cell or a sperm receives mutated DNA and the two are combined during mating, the error in the DNA will be propagated throughout the new organism, including its germ cells. Again, in the majority of cases, there is no real problem, because the error will often fall in a location of the DNA that is not expressed or is inconsequential for other reasons. In the jargon of science, the genotype (the genome) is altered, but one does not observe a change in the phenotype (the internal and external appearance) of the organism.

If a mutation does cause a change in phenotype, chances are the effect is negative, rather than positive. Why is that? Imagine that just one word were randomly replaced in a Shakespeare sonnet. How likely is it that this random alteration would better the Bard? Natural selection has led to organisms that are in some sense optimal, or at least better than their competitors. Changing one feature of an organism in some random manner is not likely to lead to an improved organism. Yet, such random changes are the drivers of evolution, which over the millennia have weeded out the weak and created us humans, which some allege to be the optimized culmination of sophistication. At any rate, evolution has successively been changing species through a process that is slow but genuinely stochastic.

Evolution is a long-term test for the competitiveness of species. It encompasses all cellular, organismal and population processes, and therefore poses challenging questions for all subspecialties of biology. It spans many levels and time scales, and is driven by systems whose components are again nested systems, which are able to adapt at several levels of organization. The attempt to capture some of the functionality and interactions of these systems has been called evolutionary systems biology. This new field combines mathematical modeling with population genetics and experimental approaches of biology to study the consequences of changes in these interactions due to natural selection.

Evolution is driven by random events. In addition to copying errors during cell division, localized alterations of DNA or RNA can be caused by environmental factors, such as radiation or toxicants, and presumably by an unhealthy lifestyle. Mutations can also occur as rearrangements of chromosomes, deletions of longer stretches of DNA, or insertions of DNA that might originate in viruses. Duplications of genes may cause problems, as we discussed before, but they also provide the raw material for the creation of "new" genes. A beautiful example is the vertebrate eye, which has light-sensing structures coded by several different genes for color and night vision. Intriguingly, these genes evolved through gene duplication from the same ancestor gene.

According to our present understanding, about 30 percent of our DNA consists of so-called transposable elements, which are DNA stretches that can move to different places in the genetic sequence, where they insert themselves. Some percentage of these are assumed to have their origin in viruses that invaded our ancestors a long time ago. While the invasion of viruses often results in disease, transposable elements are responsible for numerous functions in higher animals, and even more so in cultivated plants. Another mechanism for moving DNA is "horizontal gene transfer," a term that describes a process by which DNA can move from one organism to another outside cell division and inheritance. Among many positive and negative features, this process is responsible for the sharing of antibiotic resistance DNA among bacteria, which has become a daunting challenge for society.

Together, these random processes lead to phenotypes with different levels of fitness, which are heritable into future generations. Sometimes the fitness of the offspring is higher, sometimes lower. Darwin proffered that each population would engage in a "struggle for existence," in which the phenotypes with lower fitness would disappear. While this conclusion is generally considered correct, caution is needed, because fitness is not absolute, but depends greatly on the environment. Imagine a plant with a gene for drought resistance. Under moderately wet conditions, this plant might not do well in a competition with a different phenotype, so that its fitness is deemed sub-optimal. However, if the environment becomes more arid, the fitness of this plant shoots up, and the phenotype will outcompete others.

It actually appears that most if not all populations, from microbes to higher mammals, contain a fair number of phenotypes that are not optimal in the present surroundings, but may become the rescuers of the species should the environment change. Phenotypes usually change over long periods of time, and new species emerge, sometimes driving existing species to local or even global extinction.

Evolution does not happen in a vacuum. It does not alter one species in isolation; rather, the entire ecosystem evolves over time, as new environmental conditions or new species can put strong selective pressures on other species. For a systems biologist, evolution means that the steady state of the system changes, maybe slowly, but noticeably, along with a newly distributed profile of variables. There are even "fast" examples of such evolutionary changes, such as the bacterial populations in the lungs of a cystic fibrosis patient, which change in composition throughout the patient's lifetime and sometimes even become replaced by fungi. The re-introduction of wolves altered the river ecosystems in Yellowstone National Park.

We certainly do not truly understand the complex processes leading to long-term evolution, but they do not appear to operate in a gradual manner. Instead, the theory of "punctuated equilibria" suggests, based on fossil records, that many periods of little change in species lasted for a long time, but were suddenly disrupted by short, strong bursts in the emergence of many new species, which may have been driven by drastic environmental changes.

It is widely accepted as true that the processes governing these bursts, as well as the phases in-between, are predominantly stochastic. We are thus faced with a yin and yang, where each organism possesses regulatory programs that deterministically control development and functionality. At the same time, the populations of these same organisms constantly try out new phenotypes, most of which do not succeed and disappear. Nonetheless, a member of the population that is different from all others, such as a carrier of sickle cell anemia, may come to the rescue of the entire species when the species faces new conditions. This sequence of outsiders over time leads to new steps in the long-term evolution toward superior beings. Business as usual seems to be a good default strategy in daily life. However, once in a while, it is time to change from the current version 1.0 to a superior version 1.01, 1.1 or even 2.0.

14

Can't we all get along?

It has almost become a cliché. One African-American, one Asian, one Caucasian, one Hispanic, one Mid-Eastern and one Native American child are all playing cheerfully together, while advertising a new gadget or the best detergent of all time. And, of course, the group consists of three girls and three boys. The picture displays an idealized world we all want to see: peaceful, harmonic diversity. We are not just all equal before the law; if we set our mind to it, we can all get along just fine, in spite of our differences. In fact, we believe that diversity expands and enriches what we as a community can accomplish.

In nature, diversity takes on a dimension of an entirely different magnitude, and a few little *Homo sapiens* of different hues playing together are just that: child's play. A long-term investigation of a lake in Wisconsin identified roughly 10,000 different species of bacteria, all living together in what modern lingo calls a metapopulation or a microbiome. According to some estimates, a single liter of ocean water can contain 10 billion viruses; that is more than one virus per person in the entire world. These populations of populations can be found in essentially any place where life can thrive. They live in oceans, lakes, and rivers, below and above ground, as well as in and on plants and animals. A single gram of soil can contain enough microbial DNA to stretch almost 1,000 miles. Some microbial metapopulations form biofilms, which are thin layers on moist surfaces and in all kinds of pipes, sewage systems, and water tubing that one might find in less than sanitary dentists' offices, and even on our teeth. All in all, it has been estimated that 5×10^{30} prokaryotic cells inhabit the earth – that is a 5 with 30 zeroes – and very few live in a monoculture.

Whether we are looking at the macro- or micro-world, no organism thrives in isolation, and the coexistence of many species indicates that complex communities have evolved together in a dynamic and adaptive combination of genomic and ecological processes. The human body is no exception. It serves as the host to an estimated 100 trillion microbes, a number that corresponds to roughly 10 times the total number of human cells in the body. Particularly fertile is the gut, where the density of microorganisms can reach 1 trillion cells per milliliter. These numbers are enormous, but bacteria are so small that the gut microbiome makes up less than 2 percent of a human's body mass. As in lakes and soils, the microbes in and on the human body are very diverse, and experts believe that they belong to more than 10,000 different species. Gut microbes are mostly bacteria, but also include archaea, viruses, yeasts and other fungi, and a few unicellular protozoa. The most famous inhabitant of about every human gut is the bacterium *E. coli*, which is usually harmless, although we know full well that it can also come in strains that cause severe diarrhea. The composition of the microbial communities in or on our body is distinct for every niche. Even different regions within our mouth and throat host their specific populations of microorganisms, and the differences in microbial communities living on the skin of the scalp, the arms, and the armpits are enormous.

An intriguing aspect of microbial metapopulations is the following. We know that competition in macro-ecological systems leads to one or a very few dominant species for each niche. This observation is so prevalent that it is actually referred to as a law: the law of competitive exclusion. As an example, only one top predator species typically survives in a given area over a long period of time. Contrast this to microbiomes. Here, thousands of different and often very closely related species, and millions of individual organisms within these species, live in the same physical space and are exposed to the same environmental factors. They compete for the same resources but, at the same time, coexist and often depend on each other, for instance by using the waste products of other species as nutrients. They may assume distinct and complementary physiological roles, up to a point where some species do not survive without the others. Sometimes the coexisting species even demonstrate division of labor, where particular species, relying on others, cease to perform basic functions but in turn are responsible for another repertoire of metabolic and physiological activities. Thus, according to today's understanding, the competition within microbiomes is different from what we know from shared macro-environments, and a typical microbial metapopulation contains so many dependencies that it can rightfully be considered a single complex dynamic system. Adding to this complexity is the fact that microbial environments frequently fluctuate so much and so quickly that the populations never reach a steady state as is assumed in the law of competitive exclusion.

There is no doubt that metapopulations are complex, dynamic systems. They are therefore of obvious interest in systems biology. Metapopulations exhibit all the features of complexity discussed elsewhere by involving very many heterogeneous constituents, spanning several organizational, spatial and temporal scales, and collectively executing complicated tasks. They are also enormously adaptive. As a case in point, our gut microbiome usually survives antibiotic treatments, and even though it may seem that the preparation for a colonoscopy leads to perfectly clean intestines, a portion of the microbiome persists, and the metapopulation, in all its diversity, is usually back and functional within a few days. If a pollutant enters a lake, the composition of the microbial population may change drastically, but given sufficient time, the bugs tend to bounce back, as we have witnessed during the natural recovery of oceanic ecosystems after oil spills.

It is easy to fathom that research on microbiomes is complicated and requires long-term efforts from many fields of targeted, hypothesis-driven experimental biology, which are to be combined with exploratory –omics methods, machine learning, bioinformatics, and systems biology. Due to the fact that metapopulations are relatively young discoveries, comprehensive research strategies have not yet crystallized, but it is clear that intensive investigation, new ideas, and effective experimental and computational technologies are needed to understand the dynamics of these systems.

Any systematic analysis of metapopulations encounters numerous novel challenges, beginning with the fact that the vast majority of species in these populations are not known. It is not even clear anymore how one should define a species within these large metapopulations. Relatively new, sophisticated genomics methods permit researchers to classify the microbes into novel groupings, called Operational Taxonomic Units, or OTUs, but these are not definite either, because microbes easily swap genes with each other. And what the specific roles of any of these OTUs are is simply unknown in the majority of cases. To make matters even more challenging, one cannot culture most of the detected OTUs in the lab, and even if this is possible in some cases, it is not clear what the experimental results really mean. After all, the artificial lab environment is vastly different than in nature, especially because their compatriots from other OTUs are missing. It is also well known that the bugs have an enormous capacity to adapt to this artificial environment, and possibly assume roles and functions they seldom need or exert in nature.

The first order of business in characterizing microbiomes is thus to identify who is there. Most of the current approaches toward this identification focus on DNA, although other targets, such as proteins and metabolites, are also being explored. The most popular DNA-based strategy is currently "shotgun metagenomics." It calls for collecting the entire DNA of a metapopulation and does not even attempt to distinguish which genes belong to which organism. Instead, it focuses specifically on one gene sequence that encodes a ribosomal RNA molecule called 16S-rRNA. The corresponding 16S-rDNA sequence is especially useful for classification because all organisms have ribosomes and the ribosomal RNA varies slightly from species to species. The rationale of shotgun metagenomics is thus to sequence all 16S-rDNAs; because these sequences vary slightly by OTUs, the collective measurement yields the number of different OTUs, and the OTU-based classification becomes a representation of the population. In most cases, an OTU cannot be equated with a particular species, but it is often possible to associate it at least with a bacterial family, genus or order. A healthy soil or lake environment can contain 20,000 OTUs or more.

An interesting and so far unanswered question in this context is whether a microbial metapopulation, such as the human microbiome, has a "core" that is identical among most – if not all – humans. It is known that the human microbiome varies drastically from person to person. However, if one could establish a "normal core," it would be much easier to characterize microbiomes associated with diseases. One could also think about manipulating an abnormal microbiome toward normalcy; for example, by means of "infecting" the gut of a patient suffering from irritable bowel disease with missing species of the normal core. The Human Microbiome Project, a feasibility study dedicated to these questions and funded by the National Institutes of Health in 2008, has already led to the establishment of a reference database containing microbial 16S-rRNA sequences from different body sites, including skin, lower intestine, mouth, nose, and vagina, of several hundred volunteers. Projects like the Canadian Microbiome Initiative and the International Human Microbiome Consortium have similar goals.

While genes are of obvious importance, there is ultimately no way around the issue of function: which sets of species do, or potentially could, accomplish the various tasks a metapopulation requires for survival? Is it possible to establish classes of microbial species for each task? If so, is it feasible to mix and match representatives from the different classes to compose viable microbiomes? If the species collectively sustain a healthy "meta-proteome" or "meta-metabolism," does it matter which member of the microbiome exactly catalyzes which biochemical reactions or performs which functions?

Somewhat related yet different is the issue of species interactions and population dynamics. How does each species within a metapopulation affect the others? Which other species does it inhibit and which does it influence in a positive manner? If we could accurately characterize these interactions, could we use computer simulations to predict the consequences of a bloom or the disappearance of a given species? Is it possible to develop the capacity to use biological and computational modeling techniques to restore a microbiome that is perturbed by disease or environmental events?

Finally, each metapopulation interacts with its surroundings, which most certainly need to be taken into account. Our human body is sure to react to medical treatments affecting its microbiome. And even if these treatments follow the very best intentions, the body is likely to launch a host of physiological adaptations, including an immune response. As a consequence, curing an ailment like inflammation of the colon cannot necessarily be restricted to the gut microbiome but will also have to account for the host responses. If so, computational analyses and predictions must capture a much more complex disease system that spans multiple organizational, spatial, and temporal scales.

Even without considering the host or environment, any rational manipulation of microbial metapopulations requires that we acquire a solid understanding of the features of these complicated dynamic and adaptive systems. If we succeeded with this prerequisite task, the potential rewards could be enormous, because microbiomes directly or indirectly affect our own health, our local and global food and water supplies, and the health of our environments.

For many people, bacteria have acquired a bad reputation, and we are encouraged to keep or wash the "germs" away from our bodies or to eradicate them with antibiotics. After all, they cause diseases and make our naturally odorless sweat smell bad. The truth is that most bacterial species are harmless, and many are beneficial or outright mandatory for our well-being. They live in harmony with their hosts, including our own bodies, and provide numerous services. Skin bacteria protect against disease-causing species, prevent the loss of moisture, and help regulate our body temperature. The gut microbiome generates important compounds, such as vitamin K_2 and enzymes that we cannot synthesize ourselves but that are needed for the digestion of some foods. Maybe not a pleasant thought, but koala babies have to eat their mother's feces, just to acquire the microbiome needed to digest eucalyptus leaves.

The gut microbiome also prevents the growth of harmful species and produces hormones that guide the storage of fats. It is almost certain that a healthy gut microbiome is responsible for our general well-being and that significant deviations from its normal composition can be associated with a number of diseases, which range from diarrhea to inflammatory bowel disease and possibly to obesity, diabetes, and several autoimmune diseases. While the functional connections between microbiomes and diseases are not fully established yet, genomic analyses of the bacterial gut floras of obese individuals and of children with autoimmune diseases have demonstrated reduced species diversity and other distinct differences in the compositions of their microbial communities.

It is even being speculated that our gut microbiome may be capable of modifying the production of neurotransmitters, which are directly associated with the proper functioning of the brain. If so, controlling the microbiome could be a means of addressing neurochemical imbalances that are found in diseases like schizophrenia, depression, anxiety, and bipolar disorder. The gut microbiome also influences antibiotic resistance, which is becoming an increasingly difficult issue, because different bacterial species can exchange DNA through the process of horizontal gene transfer.

The importance of microbial metapopulations is not restricted to humans, of course. It has been known for a long time that bacteria colonize plants and provide beneficial and indeed necessary services. In particular, root bacteria are essential for the fixation of nitrogen, enhance the uptake of minerals like phosphorus, and convert these minerals into a chemical form the plant can use. Without ants and termites, the world would suffocate under a layer of undigested plant products. However, it is not the ants and termites themselves, but complex microbial communities in their digestive systems, that convert woody materials into food and energy. In a similar manner, digestion and rumination in cows and sheep depend critically on microbes.

So, what would it take to understand the function of microbiomes in sufficient detail to permit reliable manipulations? The list is long, which indicates that it will take some while before research findings can be translated into disease treatments or pollution control. A first class of research targets is related to the question of whether each type of microbiome contains a common core: namely, which species are truly needed? This question may sound harmless, but is actually quite loaded, because it is likely that any "core" might do just fine under normal circumstances. However, as soon as the environment imposes changes, the core might no longer be sufficient. Imagine, for instance, a lake with its several thousand microbial and viral species. Under normal conditions, just a few dozen species may constitute 95 to 99 percent of the total mass. However, given a strong perturbation, such as a pollutant, some dominant species will decline or perish, whereas some of the very rare species might step forward, proliferate profusely, and break down the foreign compounds. Thus, these rare species may only be needed for emergencies, while they are essentially invisible at most times. Pursuing this line of thought further, it is easy to imagine that different combinations of rare species can become the rescuers of the metapopulation under different perturbed scenarios. With respect to these types of scenario, the human gut microbiome seems actually easier to address than an ecological system, because it is somewhat buffered by the body, for instance with respect to temperature and pH, and because the acid in the stomach prevents many potential intruders from entering the intestines. Furthermore, one will be able to focus on one disease at a time and study variations in the associated microbiomes.

An intriguing issue is how it is possible that simple microbes are capable of establishing functional biofilms or coordinating collective responses to environmental perturbations. A key to this coordination is the phenomenon of quorum sensing, with which microbes signal their presence to each other and which already came up in Chapter 11. By emitting chemicals into the environment that are received by others and trigger the further emission of these same chemicals, the population can sense when a quorum is present. Usually, bacteria of the same species respond to these signaling molecules, but the mechanism may also be effective in metapopulations. As a pertinent example for quorum sensing, virulent pathogens, such as the dreaded *Pseudomonas aeruginosa* and *Staphylococcus aureus*, can use this sensing mechanism to form biofilms, which may ultimately harbor different species and are very resistant to drug treatments. If we would understand the dynamics and regulation of quorum sensing processes in sufficient detail, we could possibly use the process as a weapon against undesired bacterial phenomena and, in particular, design strategies to disrupt biofilm formation. One approach to accomplishing this task could be the model-based rewiring of the signal emission and reception pathways with methods of synthetic biology, which could possibly prevent the formation of biofilms and thereby decrease the resistance of the bugs to antibiotics.

If left to their own devices, many microbial species can get along and coexist, even if they live in the same environments and compete for the same resources. We are just starting to appreciate this synergistic diversity, but it will take some time before we truly understand it. As in other areas of systems biology, the tasks before us will require the collection of a lot of information. Large datasets will be assembled, and with these data will come noise and spurious results. Machine learning and bioinformatics will be employed to sift through the floods of data and help with the formulation of hypotheses. Modeling and simulation will help integrate data, analyze hypotheses, and possibly predict means of targeted microbiome manipulations and interventions. The future of this field is probably even more complex than we imagine today. But it is also very exciting, and the potential rewards are enormous and difficult to fathom. They will have implications for health and disease, food and clean water availability, energy production, and the stewardship of thriving and sustainable environments.

15

Love thyself and fight all others

No, it is not a misprint. And no, that command is certainly not cited from religious scripture or a guide to ethical living. It is a top imperative that the human body, like every other organism, must obey while dealing with the rough and tumble world, where rolling out the welcome mat to microbial or other visitors and invaders is a risky proposition. Not that all visitors are bad. There is strong suspicion that our mitochondria, which manage the energy needs of every single cell, were once free-living microbes. Also, it is not quite clear when and how it happened, but over 40 percent of our very own DNA has its origin in so-called retrotransposons, which most probably were parts of viruses that became integrated into our genome and gave us genes that we did not possess before but that are apparently useful. Finally, the body has learned not only to tolerate but to benefit greatly from complex, diverse, and distinct microbial communities that live on the skin, in the gut, and in all cavities that are connected to the outside world: the mouth, nose, ears, anus, urinary tract, and vagina.

Notwithstanding these beneficial exceptions, most encounters with foreign invaders are potentially a matter of life and death, and every higher organism must therefore take this threat very seriously. And because of what's at stake, the immune system defending our body could hardly be more complicated in its structure and function. It is truly a multi-faceted system of systems that covers every level of biological functionality, has lots of inbuilt redundancy, and is very robust, amazingly adaptive, and able to respond to threats never before encountered. It is a paradigm of what systems biology must learn to address.

In fact, while anthropocentric comparisons often fall flat, it does not take much imagination to identify a number of parallels between our body's immune system and a typical military defense apparatus or police force. There are fortifications and fences, stationary and mobile patrol units, vigilant controls at all possible entryways, and means of expelling intruders before they reach critical locales; there is sophisticated information sharing through alternative channels, fingerprinting of hostile agents, and a permanent record of previously identified offenders; there is even corruption among those who once used to be good guys, and there are sleeper cells of enemies that can stay dormant for months or years before they emerge and wreak havoc.

Many intruders threatening the body are potentially fatal if they reach the bloodstream, which they then use as a very efficient distribution network throughout the body. In recognition of this unyielding danger, the body seldom takes chances and, silently but efficiently and unscrupulously, makes most intruders disappear. Not to be outsmarted, some bugs can go into hiding or form spores with tough skins, only to reemerge when the host body is weak. Other bugs rapidly change their outside features, to which the immune system responds, thereby staying one step ahead of it. *Plasmodium falciparum*, the unicellular pathogen responsible for malaria, invades a red blood cell and causes it to secrete a substance at its outer surface that makes the red blood cell stick to the wall of small veins, thereby evading circulation and preventing the spleen from destroying it. Some cancers can dispense microcysts of a few cells, which settle in some organ beyond the reach of the immune system, and can turn into new tumors years later. Thus, the arms race is on and will presumably never end.

If one needed convincing that the body's defense system is complicated, a quick check on the *PubMed* website might be instructive. *PubMed* is a large scientific database, maintained by the US National Library of Medicine – National Institutes of Health, which stores most scientific publications in the fields of biology and medicine. There, a search for the keyword "immunology" yields 50,392 hits for the year 2013 alone. That is about one new paper every ten minutes, without ceasing!

Our body has to be prepared to face two very different types of attacker, which hail from the macroscopic and microscopic worlds. In the macro-world we need to flee or face bears, wolves, snakes, mosquitoes, spiders, and the like, and we have developed complex physiological and psychological fight-or-flight strategies to deal with them. This chapter will not discuss these threats, but instead focus on the often more hideous microscopic pathogens, bacteria and viruses that bombard us in the thousands every day and that we cannot even detect with our senses, until they cause symptoms.

The first line of defense against these types of invader is most intuitive: keep them out. Build fortification walls, secure all gates, shoot arrows and pour boiling oil on those trying to scale the walls. Indeed, for a microbe to break through a mammal's protective layers is a great accomplishment, because we are equipped with: a thick skin that makes it essentially impossible for microbes to pass; beneficial bacteria on the skin that discourage the growth of pathogens by secreting metabolites that are poisonous to most other microbial species; sweat, spit, and tears that contain lysozymes, which are enzymes that can dissolve bacterial cell walls; an oral cavity covered with microbe-trapping mucus that moreover contains antimicrobial agents like antibodies and lysozymes; mucosal surfaces that are heavily colonized by bacteria and can even contain viruses, called bacteriophages, which specifically attack unwanted bacteria; protective hair in our nostrils; and wax in our ear canals that is almost impenetrable for microbes and inhibits their growth. Microbes in the respiratory tract get stuck in the mucus-covered trachea and are eliminated through tiny hairs that propel particles upward, followed by swallowing, coughing or sneezing, or through enzymes and the action of white blood cells. We might find it strange, if not unsanitary or outright dirty, that wastewater rushes through our genitals several times a day. Yet, the vagina and penis are possible portals for invaders, and physically flushing them out regularly with a fluid that is not conducive to bacterial growth makes perfect use of an otherwise unwanted substance. There is one situation where we really cannot avoid the influx of microorganisms, namely, when they come as stowaways traveling on food. The treatment of these unwanted visitors is brutal: we produce such a highly corrosive stomach acid that it can dissolve iron nails. And if bugs survive the acidity of the stomach, they move on to the opposite challenge: an alkaline milieu in the lower intestines, where they are moreover attacked by lysozymes and bile.

However effective these precautions may be, some bugs regularly make it through the formidable initial defenses, and the body must be prepared to deal with them. Indeed, it seems that no efforts are spared in the sophisticated, well-coordinated defense system of highly specialized mechanisms. Collectively these mechanisms battle everything foreign, from proteins and some other small and large molecules to viruses, bacteria, single-celled pathogens, multicellular parasites like worms, and even the body's own cells if they happen to go rogue and must be neutralized quickly to prevent problems.

A great challenge for these mechanisms is the determination of what is good or bad, what is acceptable, and what must be fought. For instance, the body usually responds even to a single foreign cell, yet a pregnant woman tolerates an embryo that is partially foreign as it contains proteins produced from the father's DNA. Normal cells and cancer cells are identical in most aspects, and while they ultimately behave differently, it is very difficult to find distinguishing physical features. Thus, an effective system is needed that allows a very precise discrimination between self and foreign and between healthy tissue and cancer. Given the complexity of the task at hand, it is hardly surprising that there is always a chance that something goes wrong, one way or the other. If the body is too lax, we suffer the consequences of cancer or of immunodeficiency as we know it from AIDS. But if the body is slightly too stringent, we suffer from allergies, or the body may even start attacking its own healthy tissues. If so, big trouble is brewing, as examples like multiple sclerosis, lupus, Lou Gehrig's disease, and many other autoimmune diseases demonstrate. Clearly, the defense systems must not only be highly efficient, but also very well balanced and tightly controlled.

It is utterly impossible to describe the immune system comprehensively in a few pages. Nevertheless, it is instructive to highlight some of the amazing subsystems as they complement each other at different functional scales in their common effort to protect the body. Even at this coarse level of description, there are many key players, some of which are probably familiar, while others are not. The reason for mentioning them is to offer a glimpse into the enormous complexity of the immune system and its intricate interactions.

Defense is needed at all levels, even inside individual cells. Critical situations of this type arise when the body turns against itself by generating faulty molecules. Of course, there is a constant reshuffling and rebuilding of molecules whether a cell is healthy or not. Through the action of enzymes, most of these molecules are readily converted, rearranged, disassembled or reassembled into different forms that are needed somewhere in the cell. Notable exceptions are proteins and DNA, which are both large and rather stable molecules with longer life-times. Importantly, structural flaws in either of them can have dire consequences for the cell's inner workings. Even slightly altered DNA may lead to messenger RNA that encodes a wrong protein. Wrong proteins can function abnormally and have been associated with many diseases, including Alzheimer's, Parkinson's, and mad cow disease.

To avoid problems caused by faulty DNA, higher cells, and even bacteria, have developed sophisticated proofreading mechanisms that check all brand new DNA while it is being generated during cell divisions. Just to make sure, a second proofreading step is in place when the messenger RNA is translated into protein. While this internal quality control is very efficient, viruses regularly manage to get their DNA into bacteria or higher host cells, which later replicate the viral DNA. Sometimes the host notices, sometimes not. By easily getting in and out of bacteria, for instance, viruses can transport their own DNA, as well as bacterial DNA, into other bacteria, thereby sometimes making the new hosts resistant to drugs.

Foreign or faulty proteins can be defused, disassembled, and recycled with a fascinating intracellular machine called the proteasome. This machine consists of a complex of proteins that form a barrel with top and bottom lids. Before a protein is degraded, a specific set of other proteins recognizes it as old, faulty or wrongly folded and tags it by attaching to it several copies of yet another protein, called ubiquitin. As the name suggests, ubiquitin can be found in essentially all tissues of higher cells, which indicates how very important it is. The tagging with at least four ubiquitin molecules signals to the proteasome that the protein is to be unfolded and taken into the barrel, where it is cut into small pieces that may be recycled. Proteins that are tagged and then disassembled by the proteasome may also come from foreign invaders, although most intracellular pathogens are destroyed by a specialized protein complex for this purpose, the immunoproteasome. Several research studies have emphasized the need to have the proteasome activities very well balanced and regulated, because both increased and decreased proteasome activities can lead to neurological problems, and proteasome hyperactivity seems to be linked to autoimmune diseases such as rheumatoid arthritis. Much has been written about proteasome activities, and studying their details makes it abundantly clear that the protein degradation process in itself is a very complex and highly regulated system with many components.

While some problems within a cell can be ameliorated, others are so severe that the cell commits suicide for the greater good. This process, called apoptosis, may be triggered by various stresses, including some viral infections. Numerous proteins are involved in the suicide response, and they again cooperate as a finely tuned system. Not all details leading to the suicidal decision are known, but ultimately the apoptotic proteins cause a debilitating swelling of the cell's mitochondria, the nucleus becomes disintegrated, and the cell is neatly disassembled into disposable units. And all this happens in a strictly controlled manner.

In many cases, apoptosis is triggered directly by stress signals sent by cells of the immune system that sense a problem, such as a viral or bacterial intruder. These signals are received by specific receptor proteins on the cell membrane. When a signaling molecule binds to one of these receptors, it forms what is called a death-inducing signaling complex that consists of several proteins and triggers apoptosis by means of caspases, enzymes that disassemble proteins. In addition to receptor binding, a virus can trigger apoptosis through the activation of signaling proteins or even the expression of viral proteins that bind to proteins on the outer surface of the invaded cell. The immune system recognizes these attached proteins and induces suicide. Apoptosis is swift, and the molecules on the surface of the dying cell mark the cell for phagocytosis, which means that a white blood cell comes and devours it. White blood cells do not only eat dying cells, they also eliminate pathogens. Among the phagocytes, neutrophils are the first responders, while longer-term responses rest with large macrophages. The name is appropriate, as it is Greek for "big eaters."

Neutrophils are the most abundant white blood cells. They are essential for the innate immune system, patrolling the bloodstream and serving as first responders to chemical stress signals. Neutrophils can enter other tissues through the blood vessel walls and reach the location of a bacterial infection or inflammation within minutes. If a neutrophil encounters an invader, it wraps its cell membrane around it and engulfs it. Once the enclosure is complete, it becomes a small bubble inside the cell with the pathogen inside. This bubble connects to a special cell organelle that contains digestive enzymes and toxic, oxygen-containing peroxides. The enzymes convert the peroxides into small amounts of hypochlorite, which consists of oxygen and chlorine. Hypochlorite is the active compound in household bleach and is so potent that it dissolves the pathogen. Oxygen, without which we cannot live, is used as a powerful killing agent.

Neutrophils are short-lived, while their close cousins, the macrophages, can live for several months. Macrophages derive from monocytes that are produced in the bone marrow and circulate in the blood, from where they can migrate through the blood vessel wall into surrounding tissues, when they are attracted by a chemical SOS signal. Within the tissues, the monocytes turn into macrophages. Some of the macrophages always remain close to one location in a tissue while others are on patrol for foreign particles. They move around in the blood like amoebas, chasing and devouring bacteria and other invaders and killing them with the aggressive molecule nitric oxide. Macrophages are very flexible and can stretch to reach quite distant pathogens. In addition to pathogens, they can rid the blood of debris from dead cells.

Recent research has shown that some monocytes enter various tissues of the body during embryonic development. These monocytes proliferate and may later turn into macrophages if necessary. Of enormous importance among these are macrophages in the brain, because monocytes or macrophages of the adult organism cannot cross the blood–brain barrier. Thus, the embryonic monocytes become the first-line attack dogs against microbes that in rare cases manage to invade the brain or the spinal fluid.

In addition to neutrophils and macrophages, the immune system relies strongly on T-cell lymphocytes. These white blood cells come in several varieties, including CD4+ T helper cells, CD8+ cytotoxic T-cells, and regulatory suppressor T-cells. The helper cells release small proteins to activate other immune cells, including cytotoxic T-cells and macrophages, and are crucial for the secretion of different types of antibodies by yet another white blood cell type: B-cells. Cytotoxic T-cells can kill cells invaded by pathogens, block the transcription of viral RNA with interferon, and release the protein perforin, which makes holes in the membranes of cells that are tagged for destruction. Water and ions enter through these holes and cause the cells to burst. This situation is quite amazing, because the cytotoxic

T-cells actually have to turn on the cells of the organism to which they belong and destroy them. Another important cell type that specializes in destroying virus-infected and cancer cells upon contact is aptly called the natural killer cell.

Of course, it is not sufficient just to start a response, there must also be an organized finish. In other words, the immune response must eventually stop its attack in order to avoid exhaustion or undesirable feedbacks and long-term side effects. The immune system uses for this purpose special suppressor T-cells that ramp down the immune response once pathogens, pathogen-infected host cells, or cancer cells have been removed.

The signaling molecules that attract neutrophils and macrophages are proteins, called cytokines. When tissue is damaged or inflamed, the production of cytokines can be increased up to 1,000-fold. The cytokines are sent out by the tissue, and also by neutrophils and macrophages that had already arrived at the site of the problem. They are modulated by the so-called complement system, which consists of a group of proteins that circulate in the blood in an inactive form. Some of these proteins are protein-cutting enzymes that, when stimulated, cleave other proteins to release the cytokines, an amplifying process that leads to a cascade of further cytokine production.

It is well known in systems analysis that such a positive feedback mechanism can get out of hand. Indeed, inflammation can sometimes lead to such an enormous amplification that one talks about a cytokine storm, which in severe cases can be fatal. The complement proteins also attract more phagocytic cells, cause pathogens to stick together, kill bacterial cells by dissolving their cell walls, and stimulate the release of histamine, which increases the ability of monocytes to cross the blood vessel wall and enter infected tissues. White blood cells produce an additional set of signals in the form of leukotrienes, which inform neighboring cells of inflammation events. By doing so, they contribute to the regulation of the immune response.

Macrophages and lymphocytes also produce interleukins, proteins that assist in the activation of the immune system. Other cells produce pyrogens ("fire makers"), proteins that cause fever in order to harm the pathogen's proteins. Yet other white blood cells, called eosinophils, release numerous chemical compounds that control allergic reactions and inflammation and can attack multicellular parasites, such as worms.

Mast cells and basophils contain histamine and are involved in allergic reactions, as well as in wound healing.

The hundreds of different cytokines are sometimes classified as pro-inflammatory and anti-inflammatory, but their roles are in truth often context dependent and not always entirely clear. Of particular interest are interferons that "interfere" with the transcription, and thus multiplication, of viruses inside a host cell. They also activate macrophages, make foreign antigens more attractive to lymphocytes, and increase the host's ability to resist further viral infections. Interferons were the key components of early drugs for the treatment of serious virus infections and combination cancer chemotherapies.

Many invading pathogens directly or indirectly trigger inflammation as a nonspecific defense response, and damage to host cells can lead to the local release of histamine, bradykinins, and prostaglandins. Of course, histamine is well known to anyone suffering from allergies. Bradykinins dilate blood vessels, and prostaglandins cause a rushing of fluids from the blood stream into the injured tissue, causing edema and increased blood flow, which leads to redness and a warming of the inflamed site. They also signal neutrophils to come to the location of the problem and alter the blood vessel walls in such a manner that neutrophils and monocytes can leave the blood stream quickly and enter the tissue. They can even change features of nearby nerves, thereby reducing the pain threshold.

The immune system discussed so far deals with present and imminent danger. However, this immediate and unspecific "innate" response system is not alone. It is complemented by the "adaptive" immune system, which documents all types of invader, particle, and large molecule that had previously triggered a response. The two great challenges of this system are distinguishing self from non-self and remembering past offenders for a long time. The need for both is evident, and the two are closely connected to each other.

The adaptive immune response is governed by B-lymphocytes, or B-cells for short. They reside in the bone marrow, spleen, blood, and the lymphatic system and have thousands of specialized receptors on their surfaces. The key component of each receptor is an antibody, which consists of a Y-shaped immunoglobulin protein. Each antibody recognizes a specific section of a foreign or abnormal own protein, one of the sugar-lipid compounds that are abundant on the cell walls of bacteria, or some other specific molecules. Generically, such a foreign molecule is called an antigen, a term that was created as the contraction of antibody generator.

Upon antigen-antibody binding, the B-cell proliferates in a process that leads to two cell types. One secretes many antibodies that bind and mark antigens, whereas the other is a memory cell that remembers the particular antigen that triggered the initial binding. The fascinating aspect of antibody formation is that it can recombine portions of immunoglobulins into a nearly infinite number of different arrangements, each of which responds precisely to one very narrow class of antigens. The variability is due to relatively high mutation rates in the structure of immunoglobulins, which lead to different versions of antibodies when the B-cells multiply in response to a pathogen attack. As a consequence, antibodies have the potential to distinguish thousands of antigens by their structural features.

B-cells also shed the antibody molecules into the bloodstream and tissue fluids, where they scavenge for foreign objects. When an unattached antibody binds to a matching antigen on the surface of a pathogen, the pathogen is tagged and attracts neutrophils and macrophages, which subsequently destroy the pathogen by phagocytosis. If an antigen is bound to a B-cell receptor, it is internalized by the cell and causes the production of small proteins that are presented to helper T-cells, which initiate the elimination of the antigens and their carriers. The memory T- and B-cells formed upon antigen binding can have a very long lifespan, which sometimes reaches several decades. They do not produce new antibodies, but instead retain the genetic memory to recreate always the same antibodies that had been useful before. Thus, in cases of future infections with the same pathogen, this immunological memory remembers the earlier attack and launches a very fast and vigorous immune response. This memory is the basis for vaccinations with engineered bacteria and viruses that were rendered harmless yet still carry specific antigens.

Taken together, the defense systems are incredibly effective. But if they are always ready to attack, how do they know to attack foreigners but not the body's own cells? The main responsible party is a tongue twister suitable for casual chats at cocktail parties: the Major Histocompatibility Complex, which, unsurprisingly, is usually shortened to MHC. This complex consists of a large number of different molecules on the surface of essentially all cells, with the exception of red blood cells. The MHC molecules are produced inside the cell and become associated with fragments of the cell's own proteins that had recently been degraded. Together, they are put into lipid-membrane containers and transported to the outer cell surface.

In the case of macrophages, some of these protein fragments come from ingested foreign proteins. During their development in the thymus, cytotoxic T-lymphocytes learn to recognize the body's own MHC molecules and to differentiate them from foreign antigens. If such a T-cell later recognizes foreign cell surface fragments, it causes the cell to undergo apoptosis. This ability is a key component of the immune system's capacity to defend the body effectively. At the same time, this ability becomes a severe challenge for blood transfusions and organ transplants. Particularly critical is a bone marrow transplant, where essentially the entire immune system is replaced with the immune system of another person. Unless extreme care is taken, the consequence is often a system-wide internal war between the host and its new immune system, which often ends in liver damage, damage to other organs, and, sometimes, death.

The importance of MHC is difficult to overstate. First and foremost, humans devote a whopping 240 genes to the production of MHC, including some cytokines and the complement system. That is about 1 percent of all genes. In addition, each gene can lead to several alternate proteins, and both mom's and dad's genes are transcribed, which is otherwise quite unusual. As a result, only identical twins have the same MHC. Moreover, the number of MHC surface markers within a population is huge, which greatly reduces the chances that a bug would wipe out an entire population, because some individuals are likely to respond to it. The fundamental importance of MHC seems to be reflected in yet another observation: it appears that MHC affects body odor in such a way that mice, and allegedly men, prefer mates with a very dissimilar MHC, which thus equips offspring with a vastly expanded defense.

Interestingly, the immune system is able to tolerate non-self in non-dangerous situations, while it can turn against self in dangerous situations. For example, vaccines with foreign antigens simulate an infection, but the addition of special "adjuvant," immune response-boosting molecules in the vaccine preparation prevents tissue damage and the body's signaling of danger. This situation leads to tolerance without a memory response. One could therefore argue that danger signals trump the discrimination between self and non-self as the gatekeeper to immune responses.

Amazing complexity! Two systems, in charge of innate and adaptive responses, each with numerous components and processes, cooperate and influence each other, one on a fast time scale, and the other on a much slower scale. Both systems involve cellular and humoral responses, where the latter is nothing funny but refers to the humors, the Greek terminology for the bodily juices, which carry antibodies and signaling molecules. And, of course, each cell within these systems has its own complicated genomic, metabolic, and physiological systems, and communicates with neighboring cells, and often with cells far away.

As if the immune system were not complicated enough! Now a recent proposal suggests that we should totally change our thinking about adaptive responses. This new concept puts the current adversarial view of pathogens and their struggle with the immune system into the context of what we discussed in Chapter 14. Namely, it is becoming ever clearer that our health is intimately connected to the microbiomes in and on our bodies. Thus, it may not be fair or even correct to consider every bacterium or virus as our enemy that is to be defeated at all costs. Instead, it might be much more appropriate to reframe our concept of the immune system, together with all the auxiliary systems that interact with bacteria and viruses, as a giant "microbiome management system." For sure, this system still fights most invaders, but it welcomes others if they help us digest and detoxify food, possibly allow us to digest new foods, or keep "bad" bacteria at bay. In doing so, the system might over time even add new genes with novel functions to the microbiome, which could certainly be advantageous. Passing such flexibility toward new and beneficial microorganisms from parents to a child, maybe through MHC, might even increase "hybrid vigor," a superiority of offspring often resulting from stark differences in the lineages of the parents.

For the systems biologist, the incredible complexity of the immune system and its interactions with the microbiome pose unique, exciting, and daunting challenges. After all, almost every disease somehow involves the immune system, and if a hallmark goal of systems biology is an improved understanding of diseases, their personalization, and possibly the development of disease simulators that, like flight simulators, permit the exploration of diseases and alternative interventions, then systems biology has to grapple with the many aspects of the immune system. At this point, mathematical models are relatively scarce and simple in comparison to what will be needed to elucidate the component systems and, ultimately, the immune system in its full complexity. A lot is waiting to be done!

Meanwhile, out there, or rather in there, the incessant battles continue every day. There are no good guys and bad guys. Both the invading pathogens and our bodies with their own microbiomes do what they are programmed to do. But the threat is constant, and our bodies cannot afford to let their guards down. There is seldom forgiveness in the immune system and there are few second chances. Then again, we will not be victorious forever, and many of us will succumb to an infection or inflammation, or to the rebellion of our own cells. Until then, most bugs end up obliterated, because our immune system successfully fights the life-threatening invaders in merciless self-defense.

16

A billion dollars for your thoughts!

Excuse me? That has to be a misprint! Even with super-hyperinflation, how is it possible that a thought that used to cost a penny just a few years ago now is supposed to cost a billion dollars?

Admittedly, the comparison is not quite fair. The random thought to be divulged at an unexpected point in time probably was not worth all that much. Maybe more than a penny, but certainly not hundreds of millions of dollars or more! By contrast, the billion offered today is really awarded for answers to a fundamental question of being human: how intelligent thoughts are formed, stored, and recalled. Many governmental funding agencies around the world are offering substantial amounts of research funds to figure out how our brain works. In the US alone, the National Institutes of Health, the National Science Foundation, the Department of Education, and various other public and private foundations have been supporting a wide spectrum of research topics in neuroscience for a long time. In addition to these ongoing programs, the US government announced in 2013 the launch of a brand new $100 million initiative with the goal of mapping the connections among the neurons in human and animal brains. The project is fittingly called BRAIN, or Brain Research through Advancing Innovative Neurotechnologies, and has the ambitious goal of reconstructing the activities of individual neurons and their firing patterns in brain circuits. Specifically, the BRAIN initiative is being challenged to create new technologies that can simultaneously record the activity patterns of millions of neurons with a time resolution corresponding to actual mental processes. The expectation – or at least the hope – is that an understanding of these patterns will reveal how we create and manage thoughts and memories, what cognition is, how we learn and make inferences, and how molecular perturbations can lead to changes in behavior. A better understanding might also suggest how to intervene if the brain goes awry due to accidents or degenerative disease. In similar strategic efforts, the European Union plans to spend over a billion euros on simulations that explain brain circuits on the basis of neurotransmitter molecules and their dynamics, and several institutes for research in neuroscience have sprung up in Asia during the past two decades. So, the $1 billion in the heading of this chapter is really a vast underestimate.

The nervous system poses a curious multi-scale challenge, because we certainly know much about the anatomy and functionality of the brain. We also have a decent understanding of individual neurons and their inner workings. And we know quite a bit about the connections between two neurons, and about major signaling pathways and neuronal circuits in the brain. It has even become possible to make a mouse brain translucent and to visualize three-dimensional networks of neurons with astonishing resolution and at a scale of millimeters, which in this context is huge. Yet, how the activity of the complex system of neurons leads to high-level brain function like cognition and behavior is simply not understood. The nascent field of systems neuroscience, at the intersection of neuroscience and systems biology, attempts to address this daunting issue by trying to establish functional associations between molecular biology and –omics, on the one hand, and cognitive and behavioral capabilities like thinking and decision making, on the other.

It is not surprising that humans have been fascinated with the brain since time immemorial. The Egyptians of the Old Kingdom in the third millennium BC actually got it wrong: they removed all organs as the first step of mummification, and stored the liver, lung, stomach, and intestines in special jars. They kept the heart, the alleged seat of intellect, inside the mummy. However, they did not preserve the brain, because they were apparently unsure of its function and therefore assumed that it was not needed in the afterworld. By the seventeenth century BC, people had corrected this view, because they had recognized that head trauma could lead to loss of speech and seizures, and concluded that the brain had to be important. The Greeks of the sixth and fifth centuries BC, who were usually educated in Egypt, finally recognized the brain as the location of the mind. A couple of hundred years later, Herophilus (c. 330–260 BC) and Erasistratus (c. 300–240 BC), the Greek founders of the renowned medical school in Alexandria, Egypt, and arguably the inventors of the scientific method, performed sophisticated dissection studies into the anatomy and physiology of the nervous system and the brain, and they learned a great deal from their careful experiments. Herophilus clearly distinguished the cerebrum from the cerebellum, discovered differences between motor nerves and sensory nerves, and described the optic and oculomotor nerves as critical for sight and eye movement. He also posited that the brain was not only the organ for intellect but also the soul. Building upon these insights a few centuries later, the Roman anatomist, physician, and philosopher Galen of Pergamon (AD 129–c. 210) gained much knowledge of the brain, as well as the physiology of the circulatory and respiratory systems, by dissecting monkeys, pigs, sheep, and a host of other animals. His thinking influenced Western medicine for an astounding 1,500 years.

Important insights into the brain in more recent times have often come from injuries or tumors that led to specific malfunctioning. While very unfortunate for the affected patients, these instances greatly advanced our understanding of the roles of specific regions within the brain. Specific ideas about the associations between anatomical structures and brain function were vastly enhanced in 1990, when Seiji Ogawa of the AT&T Bell Laboratories introduced the first prototype of functional magnetic resonance imaging (fMRI). The core concept behind this method is that active brain sections require more energy and thus increased blood flow. By labeling oxygen-depleted hemoglobin, fMRI assesses changes in blood flow in activated regions of the brain of a person who is performing specific tasks or mental activities. Somewhat similar in concept, although technically quite different, positron emission tomography (PET) scans visualize the use of glucose by activated brain sections. Naturally, the decades following the first fMRI and PET scans witnessed huge advances in imaging techniques, and brain scans today are capable of directly exhibiting amazingly detailed neuronal activity patterns. Thanks to these and other methods, we have amassed enormous and very detailed, high-level information about the various parts of the brain and their functionality.

The key actors of the brain are, of course, the neurons. Each neuron is an excitable cell that is capable of transmitting information through electrical or chemical signals. Like other cells, neurons have a cell body. In addition, most neurons have a single axon, which is a thin tube-shaped outgrowth that may be over one meter long and branches out only at the end. The typical neuron also possesses numerous branched dendrites that are less than a millimeter long and form tree-like structures. The transmission of information from cell to cell typically occurs at synapses, which are specific contact points between the axon of a signal-sending neuron and a dendrite of the signal-receiving neuron. The axons of peripheral nerve cells are usually insulated by specific Schwann cells, which wrap around the axon, prevent the unwanted spread of electrical charge, and even speed up signal transduction. An axon of one-meter length is insulated by about 10,000 Schwann cells. Active neurons require nutrients and quite a bit of energy, which are supplied through astrocytes. In addition to protecting and insulating brain neurons, and servicing their metabolic needs, astrocytes control the blood–brain barrier, which is the brain's ultimate guard against microbial invaders and large molecules. Astrocytes constitute by far the most abundant cell type in the brain.

Enormous research effort has been devoted to neurons, their signal transduction capabilities, and their support cells, and we believe we have a pretty solid understanding of how neuronal signaling works. The signal transduction is mostly electrical, with the electrical charge deriving from chemical processes and, in particular, the transport of charged ions in and out of cells and between different compartments within the neurons. In fact, one of the true early successes of mathematical modeling in biology was the 1952 discovery by the English physiologists and biophysicists Alan Lloyd Hodgkin and Andrew Huxley that the initiation and propagation of electrical signals in neurons can be formally represented as an electrical circuit, as we find it over and over again in electronic gadgets. Hodgkin and Huxley, who in 1963 received a Nobel Prize for this discovery, demonstrated that an electrochemical gradient corresponds to a voltage source, ion pumps can be represented as sources of electrical currents, the cell membrane acts like a capacitance, and ion channels correspond to electrical conductances. By establishing these analogies, Hodgkin and Huxley were able to analyze neuronal activities like features of well-understood electrical circuits. As the only real difference, which however is crucially important, they recognized that the neuron is able to modulate the function of its ion channels and therefore its electrical

activity. In a stroke of genius, Hodgkin and Huxley came up with a set
of three differential equations representing this modulation capacity.
Of course, science being science, several extensions and improvements
were proposed throughout the following half-century, but Hodgkin
and Huxley's analogy between a complex biological phenomenon and
an electrical circuit, and thereby a computational model, is still considered one of the greatest successes of biophysics and biomathematics.

We are not only cognizant of the electrical aspects of neuronal activity
but have also gathered considerable knowledge of the chemical aspects
of signal transmission at a synapse, which is accomplished by neurotransmitters. These brain-specific molecules are released by the signal-
emitting neurons and accepted and interpreted by receiving neurons.
The human body makes use of about 20 neurotransmitters, as well as
a number of peptides and ions that may also be used for signal transduction. The most frequent neurotransmitter is glutamate, but the lower
concentrations of the others should not imply lesser importance. For
instance, dopamine is of enormous interest, although its concentration
is minute in comparison to that of glutamate. The reason is that dopamine affects our reward system and can make us feel good, whether the
trigger is food, a pleasant sight, success or a drug like amphetamine. It
also causes substance abusers to feel lousy when the drug concentration wanes, and is associated with Parkinson's disease, whose hallmark
is an insufficient amount of dopamine in certain parts of the brain.
Generically, the signal-emitting cell synthesizes the neurotransmitter
molecules and packages them into specific intracellular vesicles that are
tiny containers made from a lipid membrane. These can be stored or
moved to the cell membrane at the synapse, where they open up and
release the neurotransmitter into the synaptic cleft, the space between
the axon of the sender and the dendrite of the receiver. With sufficient
neurotransmitters in the cleft, the receiving neuron responds in some
appropriate fashion, or ignores the signal if it is too low.

According to widely shared belief, many mental disorders like schizophrenia and depression are accompanied – or even caused – by imbalances that disturb the normal neurotransmitter profile. These imbalances
affect the proper transduction of nerve signals across synapses, which in
turn changes the dynamics of important brain circuits. Not surprisingly,
the balance or imbalance among neurotransmitters is not easy to assess,
as it is the manifestation of a complex biological system, in which the

neurotransmitters affect each other's activity and effectiveness, so that disturbances in one neurotransmitter propagate through the brain in a convoluted manner.

In recent years, chemical signal transmission at the synapse has become a common target of two groups of scientists that did not collaborate much before, namely, on the one hand, computational neuroscientists, who mostly focus on individual neurons, electrical events at the membrane, the functional connections between neurotransmitters, and the behavior of neuronal networks, and, on the other hand, systems biologists with their newly acquired arsenal of high-throughput experimental and computational –omics methods. There is even talk about a new research field of "synapse systems biology," which is quite appropriate, as the synapse indeed constitutes a very complex system: at the "presynapse" of the axon, neurotransmitters are produced, stored in vesicles, moved around, and degraded, which requires numerous metabolic and cell physiological functions; the synaptic cleft is the location of the dynamic release, diffusion, and reuptake of one or more neurotransmitters; and the "postsynapse" of the dendrite receives signals, integrates them with specialized proteins on a short timescale, and controls long-term genomic adaptations to frequently repeated strong signals, which are the hallmark of both learning and substance abuse.

Simplistically speaking, a neuron tends to receive many simultaneous input signals, some of which are activating and some inhibiting. The cell somehow integrates all these signals and in turn sends a signal, or not, to the next neuron in the signal transduction chain. This simplistic picture inspired the invention of artificial neural networks, which do just that: they consist of a first layer of "neurons" that receive many inputs, weigh them according to their importance, sum up the result, and send this result to a second layer, where the incoming signals are again weighed and summed. The output is typically one number, namely, this last computed sum. As was discussed in a different chapter, artificial neural networks can be trained to recognize complicated patterns or distinguish one pattern from another. As an example, such a network may be trained to distinguish cancer cells from normal cells.

Taking all of the above together, we believe we understand the functioning of the base unit of the brain, namely, the neuron, reasonably well. We also have a good idea of how signals are propagated across a synapse and how several signals are weighed against each other to evoke a response or to ignore the signal. Moreover, we are aware of major signal transduction routes within the brain and have quite a clear picture of what each brain section does. Alas, we cannot explain cognition or memory. Why?

Much of the complexity of the neuronal signal transduction system arises from its sheer size. According to best estimates, the human brain contains about 100 billion neurons, as well as many times more astrocytes. Each neuron has on average 10,000 synapses, for a total of roughly 100 trillion neural connections in the brain! That is over 1,000 times the number of stars in our galaxy. The new neuroscience initiatives intend to map out and document a good portion of this "connectome." The task is surely daunting. In addition, it is complicated by the fact that even our closest relatives in the animal kingdom not only have much smaller brains but also much simpler wiring patterns, so that results from animal studies have to be treated with caution. Moreover, in contrast even to our closest animal cousins, our brains possess large "association" regions with complex neuronal connectivity patterns, which are located between the regions controlling our senses and movements. It is being hypothesized that these association regions are crucial for retrieving memories and making decisions.

How quickly – or to what degree – the connectome effort will succeed is anyone's guess. The international Blue Brain Project in Geneva has already created a detailed huge-scale simulation model of a small part of the human brain cortex, yet, in spite of its size and complexity, the represented part is tiny in comparison with the entire brain. Then again, computer power has been steadily increasing, and one might expect that we will establish at least a coarse connectome in the foreseeable future. Furthermore, one might expect that the insights gained could help us identify specific differences between healthy and diseased brains.

A deeper issue than the magnitude of the task before us might actually be that cognition, language, memory, thought, the envisioning of the future, and other high-level functions of the brain clearly are emerging features. After all, it seems evident that a few neurons by themselves will not offer a comprehensive explanation, but that it is instead the enormous complexity of the neuronal system in the brain that must be responsible for these emerging, high-level capabilities. This conclusion suggests a nagging fundamental question, namely: will we ever be able to reconstruct high-level brain function from its elementary components, whether these are individual nerve cells or brain circuits, or are there principally insurmountable obstacles to explaining the emergence from the system's constituents? Mark Bedau's argument, discussed in Chapter 9, comes to mind; namely, that the aspect of "getting something for nothing" in emergence is "illegitimate magic" and that the best we might be able to achieve is to capture a weak type of emergence, which at best can be mimicked through computer simulation. Will a future extension of an initiative like the Blue Brain Project allow us to simulate a thought from a complicated neuronal network model and, if so, will we truly understand whence and how the thought emerged? If we look at flying as an analogy, it is clear that our airplanes are much faster and have incomparably wider reach and higher capacity than birds, because many of their engineering design principles are genuinely different. Will our computers think faster and better, but in a fundamentally different manner? Already the best computers can beat chess masters and certainly have a much better memory than any human. We have computer-based expert systems that correctly answer complicated questions, and small termite-inspired robots have enough "swarm intelligence" to collaborate autonomously on the construction of towers and pyramids. Lie detectors can assess the verity of our thoughts, and it seems to be only a matter of time before we have a so-called Turing machine, which is a computer system, named after the brilliant twentieth-century British computer scientist Alan Turing, which can respond to questions so intelligently that a human observer cannot tell whether the answers were given by another human or the computer.

No doubt, these are fascinating, fundamental questions, and neuroscientists, computer scientists, systems biologists, and last, but not least, the lay public will eagerly follow the progress toward deciphering the most puzzling organ in the human body. As for a particular thought right now, we may just have to pay that penny.

17

The computer will see you now...

Unsurprisingly, health and disease have always been on the minds of humans. After all, this most fundamental yin and yang of life is without doubt one of the strongest drivers of our happiness, success, and general outlook on life. The more we learn the more we realize that our physical and mental well-being are ultimately a reflection of the myriad of biochemical, electrical, and mechanical processes chugging along in our bodies every minute, day and night. Slight derailments of some processes we might not even notice, but others, just think *toothache*, are annoyingly distractive, even if 99.99 percent of our physiological machinery is working just fine. No wonder that the desire to understand the connections between molecules and a happy or not so happy life have fascinated biologists for many years.

When systems biology began to gain popularity at the beginning of this century, health and disease were portrayed as prominent targets, but with a new twist: the novel promise of personalized medicine and predictive health. Building upon this promise, the director of the National Institutes of Health, and the commissioner of the US Food and Drug Administration, laid out a road map for investments in this new field of "4P medicine," which was to become Personalized, Predictive, Preventive, Participatory. They also proposed efforts to streamline the path to new and improved treatments, which were to be custom-tailored to the individual. These treatments should dispel the old cliché of the doctor's advice to "take two pills and call me in the morning." Instead, they portrayed a vision of science-based efforts to adjust and individualize important disease treatments according to each patient's age, metabolism, and many other personal characteristics. This vision implied that government, universities and drug companies would team up to develop new, effective drugs and diagnostics so doctors would be enabled to prescribe "the right drug at the right dose at the right time."

The vision of predictive health goes one step further. The basic idea is quite simple, but its implementation will be time-consuming and require vast resources. In the first phase of this implementation, dedicated health centers will perform thousands of measurements on large cohorts of healthy individuals of various ages and repeat these measurements regularly over a time period of several decades. As one might expect, some of these individuals will eventually develop a serious chronic disease. If this happens, one will be able to sift through the sick individual's electronic medical record and try to identify where the formerly well-oiled machinery of the body began to veer off the path of health and exhibit unusual molecular features. Initially these deviations presumably did not even cause noticeable symptoms or might have been considered harmless, but they could in hindsight stand out as troublesome. The thought is that with enough such data, methods of machine learning and advanced statistics could eventually be employed to connect very early irregular events through a chain of causes and effects to the noticeable onset of the disease. For the particular individual in question, the realization that the disease started quite harmlessly several years ago would not be a real consolation. However, if enough of such retrospectives could be collected, future individuals with similar early features could greatly benefit, because one could make increasingly more reliable prognoses of an individual's future health. In fact, one could possibly make rather reliable predictions of what would happen if the individual simply went on as if nothing had happened or, alternatively, received proper treatment. Of course, such predictions will require massive amounts of data, which today do not exist.

Systems biology is expected to contribute to the field of personalized medicine and predictive health in three domains, namely: data acquisition; the formulation of hypotheses regarding risk scenarios; and the analysis and interpretation of vast amounts of information with computer models. Let's look at these three aspects in more detail.

Any discussion of predictive health needs to begin with the modern buzzword of biomarker, which simply describes any specific feature that is measured in the context of a medical exam and could potentially have a bearing on health or disease. In the olden days, a visit to the doctor included a number of tests that culminated in a report on biomarkers like blood pressure, heart rate, lipids, blood glucose levels, liver enzymes, and urine data. Typically, the doctor obtained maybe a dozen such biomarkers and combined them with patient statements such as lack of energy or abdominal pain and with her own observations, such as a yellow tint in the white of the eye or discolored fingernails. She assessed the collective results against her own experience and against the state of the art in family medicine, which typically consisted of "normal" reference intervals. Based on what was considered normal, the good doctor made a diagnosis or ordered more specialized tests, such as a chest X-ray or a comprehensive allergy test.

Thanks to the enormous advances in molecular biology and medicine, this almost quaint Norman Rockwell picture is beginning to fade into the past, making room for data acquisition with methods of –omics, imaging, and a number of other technologies that easily characterize thousands of biomarkers. For instance, it is no longer difficult to measure the expression of all of our roughly 20,000 genes, all in one experiment, and it is possible to identify tens of thousands of chemical compounds in a drop of blood serum with modern methods of mass spectrometry. Images from a CT or MRI scanner now have such exquisite resolution that they exhibit details we had never imagined to be visible without a biopsy or an autopsy just a few decades ago.

The huge amount of information obtainable from a single individual is quite a miracle but, alas, the blessing comes with a curse. Namely, each one of the thousands of data points by itself seldom tells us much about health and disease. Even if a single biomarker turns out to be important, it rarely tells the whole story. Some heavy smokers have disconcerting biomarker readings all their lives, yet live into their nineties, while others who have never smoked develop lung cancer nevertheless. Nature offers no guarantees of fairness. Some people naturally have cholesterol levels that are considered too low or too high, but they feel just fine, meaning that "their normal" is just different from the "average population normal." We might credit or blame our genes for health and disease, according to the old joke among epidemiologists that the best recipe for health is picking the right parents. However, important as our genes are, they alone do not determine our well-being, and we can easily list many other factors possibly affecting our health. Lifestyle and diet, preservatives and pesticides in food, detergents and cleaning agents, and exposure to chemicals in the environment are the obvious tip of the iceberg. To complicate matters further, these chemicals could have exerted their effects not merely in recent times, but possibly many years ago, maybe even at a time when we were infants or before we were born. Indeed, new research suggests that some exposures that our parents or even grandparents experienced might contribute positively or negatively to our well-being. Not all details of these processes are quite clear yet, but we are learning more and more about them from studies in the relatively new field of epigenetics, which studies inheritable changes that occur not through our genes but through molecules attached to our DNA that allow our genes to be turned on or off in certain situations.

It is clear that health and disease are outputs of the many complicated molecular systems governing our bodies. Our personal characteristics, such as height and muscle mass, intelligence and interests, intolerance to lactose or loud music, are convoluted consequences of unknown combinations of our genetic makeup, epigenetic markers, and lifelong interactions with other humans, animals, and the environment. Health and disease are no different, and we seldom have a clear idea which combinations of genes and exposures are good, bad or indifferent for any given individual. We cannot even say anymore what is *normal*, if thousands of genes and biomarker measurements are in play, and it is indeed very difficult to define *health* without using words like *disease* or *sick*. So, we have a novel scientific problem. In the past we never had enough data. Now we sometimes have so much information that we no longer know what to do with it.

A typical medical exam provides us with a snapshot of a part of our biomarker profile. This snapshot contains valuable information, but is often not sufficient to distinguish health from disease. A series of snapshots over time, by contrast, may tell us a lot. If a biomarker is occasionally outside the "normal" range, there may not be a problem. We all have had fevers or a bout of high blood pressure. However, certain combinations of biomarker that are frequently out of bounds may suggest that something bad is brewing. To come to such a conclusion is difficult though, because a modern biomarker profile may consist of thousands of data, and because the body has enormous self-healing powers that, often unbeknownst to us, can compensate for many deviations from the normal biomarker profile. The only way to deal with these large datasets and develop new hypotheses regarding a disease is the use of computer algorithms from the field of machine learning, which we discussed in Chapter 3.

In the context of personalized medicine and predictive health, the machine learning algorithm must be trained long before any patient can benefit from it. Specifically, the algorithm is given a large set of training data consisting of extensive biomarker profiles from many thousands of individuals, and additional information regarding which of these individuals have a certain disease. To simplify the discussion, let's invent the disease "ignoritis." The computer algorithm tries to identify patterns of biomarkers that distinguish people with ignoritis from individuals without it. Suppose 300 people in the training dataset have ignoritis and that 262 of these people have: (1) strongly elevated biomarker #1422; (2) slightly elevated biomarker #716; and (3) strongly lowered biomarker #5275. Suppose further that most of the 4,700 healthy individuals in the training set do not exhibit this combination. Then, the computer makes the deduction that this particular combination has something to do with ignoritis. The computer analysis can also provide a coarse quantification of how likely it is that the deduction is indeed correct. However, there is no explanation regarding causality or with respect to specific mechanisms or processes in the body that could potentially associate the three biomarkers with ignoritis. In fact, it might not even be clear what exactly biomarker #1422 is. It could be the elevated expression of a gene whose function we do not know. Nonetheless, if the computer prediction can be validated in a second phase of the analysis, with data from additional patients, the combination of biomarkers may be considered a valid candidate for diagnosing ignoritis. While the basic concepts of this machine learning approach are solid, it is not surprising that the task of identifying true associations between disease and novel biomarker combinations quickly becomes a humungous task, which becomes less reliable as more and more biomarkers are measured, unless huge patient databases are assembled.

If the computer spits out suspicious combinations of biomarker readings, some of them correct and some maybe not, then three questions arise. First, can we separate the wheat from the chaff? Second, if we identify certain combinations of biomarkers as statistically significant, are they necessarily causative and, if so, what do they mean? And third, if we are able to interpret these combinations, does it help us with the design of countermeasures?

These questions bring us to the third contribution of systems biology to the areas of personalized medicine and predictive health. This area is the interpretation of molecular information not within the direct context of other molecules but within the context of physiology and disease. The main tool for this type of analysis is an explanatory computer model. Such a model is distinctly different from machine learning algorithms and instead simulates the processes and mechanisms that drive a healthy body into a specific disease.

The analogy of a flight simulator might be an intuitive means for explaining the concepts behind such a model. A flight simulator is used for pilot training. It looks like the cockpit of a real airplane with all the controls, switches, and instruments, and the computer algorithm in the background very realistically simulates two types of situation. One addresses the normal life of a pilot, and allows practicing take-offs, landings, and all normal functionalities in-between. The second type portrays common or rare problems, from simple instrument failure to potentially catastrophic events requiring an emergency landing. The purpose of the flight simulator in this case is to hone the pilot's ability to test different approaches to solving these problems without the risk of crashing a real plane.

Similar to flight simulators, systems biology will eventually generate disease simulators. These simulators will be based on very detailed, mechanistic models, which, however, the user will never see. These dynamic models will typically contain hundreds of differential equations that describe how the body responds to small or large perturbations. They will also account for stochastic features, such as unusual responses to drug cocktails or unpredictable events that turn a healthy cell into a cancer cell. "The body" is modeled to correspond to an average of the "normal" population, and the parameters of the model will be set to correspond to normal ranges, as defined by clinical practice. To the user, the simulator will present a specific patient along with symptoms and initial biomarker measurements. Mathematically speaking, the computer creates this patient by replacing some of the population parameters with patient-specific parameter settings. Some of these parameters correspond to typical biomarkers, while others are associated with the simulated disease. For instance, a healthy individual should have a total cholesterol level of less than 200 milligrams per deciliter of blood. In our patient, this value might be 245. Blood pressure might be elevated from 140 over 70 to 162 over 98. Heart rate could be close to normal. And so on. The student doctor is presented with output from the simulator and asked to make a diagnosis or order further tests. To simulate new test results, the computer retrieves information from the patient's internal database and calculates with the dynamic model what the specific test would show. The simulator returns the results and asks the budding physician again to make a diagnosis and prescribe a treatment or order further tests. The simulator always obliges politely, administers the treatment to its virtual patient, and tells the physician whether and to what degree the treatment was a success. As in the case of pilot training, the simulators will cover normal cases as well as highly unusual situations that might be reminiscent of the TV show, *House*.

Disease simulators will become indispensable teaching tools. They are also likely to become the first method of diagnosis. It might be a sobering thought, but a visit to the doctor's office might resemble a car's "visit" to the garage. As with the car, the patient will be hooked up to a computer. There will probably be the friendly interface of a nurse, but all data, including the patient's electronic health history, will go directly into a computer algorithm. The algorithm will not dictate the next medical steps but instead make suggestions to the doctor or nurse practitioner, and the further into the future we look the better these suggestions will be. The doctor may still override them, but probably will not do so, maybe with the exception of very unusual cases. It is left to you, the reader, to ponder whether the doctor's hunch might yield better results than the computer program. No matter how a diagnosis is made, it will be the physicians who are performing operations, supervising treatments, and caring for their patients.

Critics may say that the thought of computer-driven diagnostics is preposterous, because the human body is simply too complicated. However, internet programs like WebMD are already quite sophisticated and will rapidly improve further. For a true individualization of diagnosis and treatment, systems biology is not yet ready. However, there is little doubt that this situation will change and eventually tip the balance toward computer-aided, personalized medicine. The amount of medical knowledge is expected to double every eight years or so henceforth. Human brain capacity, by contrast, will not do so. There will come a point when cheap tests routinely yield vast amounts of reliable bio-marker data and where doctors will no longer be able to interpret them in their heads. Meanwhile, computers will continue to get smaller and better, and mathematical modeling will eventually catch up to a point where models become reliable partners in medicine. Today, a doctor's intuition is still a great asset, but eventually disease simulators will allow the exploration even of a vague intuition.

Sometime in the not-too-distant future, a friendly receptionist or nurse may still call you in, and you may not even realize that the computer will see you then.

18

Redesigning perfect

Whether we like bugs, viruses or creepy crawlies of various stripes does not really matter. But we do have to admit that each of them has, as a species, survived for a long time. Each one of these critters has consistently been more competitive than all challengers. It has been superior to others and therefore is in some sense optimal. Each has passed the strict test of evolution by surviving in an often hostile world. We might actually be forced to admit that each one of them is perfect in its own way.

So, if a bug is perfect, can it be improved? Why should we even think about changing it? Practitioners of the new fields of synthetic biology and metabolic engineering, which could be considered prominent applications of systems biology, certainly believe that exactly that could be very rewarding. Synthetic biologists and metabolic engineers have no problem with the argument that bugs are perfect, but their rationale for trying to change them derives from the fact that perfection may be judged by rather different criteria.

A mathematician might say that the bug has to solve a "multi-objective optimization" task. It must optimize many objectives simultaneously, and these objectives are often in conflict with each other. We deal with such tasks every day. For instance, buying a winter coat involves numerous objectives that should be satisfied and, ideally, optimized. The coat must have the right insulation for the expected temperatures, we really want to like the material and color and, of course, the price must be right. Abundant experience tells us that we are seldom able to satisfy all criteria completely and that we must prioritize or compromise by weighing different aspects against each other. Even the nicest coat is out of reach if we cannot afford it.

All organisms in the real world have this type of problem throughout their lives. Above all, they must ensure that their species survives and proliferates. This overarching objective involves uncounted smaller tasks, as the organisms need to be tolerant to environmental perturbations, fight off, avoid or flee from hostile challengers, appeal to potential mates, and bring their young ones to independence. The rationale of metabolic engineering for altering organisms is that if one relieves the organisms of some of these tasks they should be able to devote energy to others. Thus, keeping the environmental conditions constantly perfect, one might "encourage" a bug to produce some valuable compound in an amount that is much higher than what it generates in the wild.

Of course, the core rationale is not entirely new. Ever since humans began to engage in agriculture, they knew that plowing, watering, fertilizing, and weeding have a good chance of improving yield. As a result, a carrot that we buy in a supermarket today has several hundred times the volume of its wild type ancestor. A turkey that was bred for Thanksgiving is sometimes so heavy that it can hardly stand on its feet; how long could it possibly survive in the wild? What's different with the new types of manipulation of organisms is the repertoire of methods that is available to today's synthetic biologists and metabolic engineers. Also, the targets are often microorganisms rather than the macroscopic plants and animals that we have been breeding for a long time.

A great leap forward in manipulating organisms began in the 1950s and 1960s with our vastly improved understanding of the details of genetics. Of course, we had known about genes and inheritance since the Austrian monk Gregor Mendel performed his experiments on peas, but it was the emerging and very effective methods of molecular biology that taught us much about the structure and regulation of genes. We deciphered transcription factor networks, and learned about different types of promoter that control the reading of the genetic code and the alternative splicing of translated mRNA stretches into different proteins. We discovered epigenetics and found out that micro-RNAs can regulate and even silence the expression of specific genes.

For a molecular biologist, these successes were fantastic academic breakthroughs, but they soon became more – namely, potential targets for manipulation. If one could splice a foreign gene into a bacterium, maybe the bacterium would use the gene and synthesize a desired product. Indeed, the biological community became really good at this particular task. We learned not only how to introduce foreign genes, but also how to control their expression. So, if we can alter one gene, we are asking now, why should we not be able to create an entirely new pathway? Why not rewire and remake an essentially new organism according to our needs and wishes? It goes without saying that such a capability would revolutionize uncounted fields, from food production to sustainable energy and novel medicines.

Amidst a lot of progress and successes in changing bugs, we came to realize that biology is not always a willing partner when it comes to remaking organisms. Foremost, there are natural limits to the capacity of the machinery that reads genes and generates proteins and metabolites. Also, these proteins and metabolites need to be maintained and controlled, thereby creating what's called "metabolic burden" in a cell that is already busy with its own tasks. As a response, foreign DNA is often ejected or disabled. We have also learned that each bug has certain preferences within the genetic code that correspond to the various amino acids, and if one ignores this codon bias, the bug is unhappy and finds ways to ignore the DNA.

The hope is that systems and synthetic biology might provide genuinely new tools to overcome some of these issues. A first target area derives from the fact that a change in a single gene or protein is usually not all that beneficial. Often three, four, five or even ten simultaneous changes are needed. As an example, imagine a complex metabolic system where the task is to increase the production of some compound we like, such as an expensive amino acid that we could use in prescription drugs or as a valuable animal feed additive. Due to the intricate regulation within the system, it is often impossible to predict which genes or enzymes should be changed, and by how much, in order to increase yield. The field of metabolic engineering has been struggling with this issue for a long time and, not surprisingly, the experimental effort grows exponentially with the complexity of the system and quickly becomes unreasonably expensive. A cheap, promising alternative in this dire situation appears to be the construction of comprehensive computational models that can help with the prediction of success or at least the weeding out of strategies that are likely to fail. Indeed, this goal of predicting how an organism responds if we introduce changes is a central theme in systems biology. The task is certainly difficult, but the strategies of systems biology are appropriate and will eventually work.

The concept of using models for metabolic engineering is easiest to illustrate with a specific example. It is well known that the sour taste of lemons derives from citric acid. This compound has many uses, not only in the food and beverage industry, but also as a cleaning agent, detergent, and rust remover. Incredible as it may sound, almost two million tons of citric acid are produced worldwide every year, half of which goes into beverages. Not surprisingly, most of this citric acid no longer comes from lemons. Instead, it is produced by a tiny black fungus, *Aspergillus niger*. Almost 100 years ago, the American food chemist James Currie received a patent for citric acid production with this fungus, and within two years, Pfizer began industrial production. In nature, the fungus does not produce huge amounts of citric acid, but metabolic engineers have been spending enormous time, effort, and money to alter the genetic makeup of the fungus, as well as the growth conditions, in order to increase yield. This effort has been very successful, but as the demand continues to climb, the question arises of whether it is even possible to increase the yield any further. More generically, one should ask whether there is an alternative to 100 years of experimental trial and error if a metabolic compound is needed in large quantities.

Systems biologists argue that both questions could be answered with complex computational models. Such models, usually consisting of large sets of differential equations, ideally represent the entire metabolic system of a suitable microorganism, including its regulation, and may be formulated as models characterizing just the steady state or the full dynamics of the system. Setting up a metabolic model is doable in principle, but the devils are, as always, in the details. The major bottleneck is currently the availability of sufficiently detailed data characterizing the kinetic features of the metabolic system. These data are needed to determine parameter values that make the model computable. Data generation and parameter estimation require a lot of effort, on both the experimental and computational sides. However, once a suitable model is established and validated, it can be analyzed incomparably more quickly than any wet-lab experiment. As a case in point, modern PCs render it possible to execute several million simulations with a metabolic model in a timespan of hours, if not minutes. And each of these simulations corresponds to one laboratory experiment!

So, simulations are cheap, but what exactly is to be simulated? Trial-and-error approaches have limited capacity, certainly in the lab but even on a computer, and the metabolic engineering community is therefore looking for effective alternatives. In particular, metabolic engineers are very interested in reliable predictions that characterize the optimal redirection of metabolites toward a desired compound, without being detrimental to the growth and well-being of the microorganism.

One strategy of assessing alternate options for improving yield is a large-scale computer simulation of very many different combinations of enzyme activities and their effects. An even better alternative is the optimization of the model with the objective of maximizing yield, while taking into account all the constraints within which the organism has to operate, including upper limits on fluxes and concentrations and a reasonable metabolic burden. In the case of citric acid, preliminary work of this type suggests that the metabolic production system can only be improved if at least six or seven enzyme activities can be altered simultaneously and in the correct amounts. This result is quite interesting because one might have thought that just searching intensely enough could reveal a silver bullet solution in the form of one key gene or enzyme. The speculated reason that several changes are necessary is the following: for over 100 years, many researchers have worked toward improving citric acid production. Chances are that they found all the "easy" solutions, which involved only one or a few alterations. However, to identify a well-coordinated, simultaneous change in six or seven reactions by trial and error is simply not very likely.

Several successful outcomes of this model-based strategy have already been reported in the literature. Examples include the microbial production of the amino acid lysine and of the protein carnitine, which are used as dietary or feed supplements. In the former case, the microbe *Corynebacterium glutamicum* was modified based on model predictions, and the altered strains now produce hundreds of thousands of tons per year, even though the wild type does not even secrete this compound. In the latter case, a dynamic model of the metabolic pathway in *E. coli* was optimized, and this model identified reactions whose alterations improved yield. Actual genetic implementation indeed led to a large increase in productivity.

As a more detailed, intriguing example, consider recent work by Friedrich Srienc. He and his laboratory set out to create strains of

E. coli that convert non-edible plant biomass into hydrogen, fatty acids or industrial ethanol. They used a systematic computational approach that started with a model of the wild type strain, which was iteratively reduced in complexity and ultimately resulted in predicting the makeup of a new, highly efficient production strain. The key to their strategy was using computational modeling to identify redundancies within the metabolic system. These redundancies allow the organism to synthesize important compounds through different pathways, which is critical for survival in changing environments. However, under favorable and tightly controlled conditions, these redundancies are not needed, so the investigators eliminated key genes in the model that were responsible for redundant pathways. The result was a strong reduction in the organism's physiological machinery and a substantial decrease in metabolic burden. Specifically, the wild type produced ethanol from the plant sugars xylose and arabinose in about 1,000 different ways, but mutations in just seven genes reduced this number to only six. The resulting strains were predicted to omit the production of hundreds of undesirable intermediate metabolites and to concentrate their efforts on the production of ethanol. The computational predictions were tested in the laboratory, and one of two strains with the reduced pathway systems produced ethanol and biomass in sufficient amounts.

Computer-based metabolic engineering of the type discussed so far occurs in a top-down fashion, which starts with a wild type strain. In parallel to this approach, systems and synthetic biologists have started at the opposite end by attempting to create small biological modules that operate as predicted by a model. This type of artificial design serves two purposes. First, it confirms, refines or refutes our understanding of how organisms function. As the famous physicist Richard Feynman said: "What I cannot create, I do not understand." Second, the ability to create modules with a predictable function obviously opens an entirely new realm of huge potential.

First steps in this direction have already been taken. Beautiful examples are mini gene circuits that were designed in collaboration between molecular, systems, and synthetic biologists. These designs are created on the drawing board as sophisticated hypotheses of what a microorganism would do if the control structure of its gene expression were changed. Instead of testing such a hypothesis in the lab, it is first cast as a mathematical model and tested for feasibility. The mathematical results are then evaluated and new hypotheses are formed until the modeling gives a green light. Only then is the model structure implemented in an actual microorganism, by introducing the designed gene system into the microorganism.

In several hallmark cases, the genetic implementation in living bacteria led indeed to the predicted results. One of the early examples was a so-called toggle switch that consists of two genes that each code for a repressor of the other. If one repressor is inhibited with a drug, the system moves from one steady state to the other, in exact agreement with the model predictions. A more complex circuit design, which was also first tested with a mathematical model, led to robust oscillation patterns, as predicted, and these oscillations were even synchronized within the microbial population. As one may expect, the sky is the limit in this new arena, and we can already see some ingenious inventions. For example, the lab of James Collins created a strain of *E. coli* that can count, and the labs of Christopher Voigt and Andrew Ellington enabled bacteria to see light and to detect edges between light and dark.

Granted, these successes were achieved with very small systems of a few genes and their regulators. So, while the first steps have been taken, we should ask: what's next? Some researchers in synthetic biology believe that the next natural step is the systematic production of simple modules that can be hooked together to create larger, more sophisticated modules. Inspired by the often observed modularity and hierarchy of biological systems, as well as the much more mundane Lego blocks, such units are called "BioBricks™." Of course, the bricks by themselves are not sufficient, as it is also necessary to develop robust and effective ways of functionally connecting them. Here, the inspiration comes from the electronics revolution, where transistors, capacitors, and other base units have been connected, stacked, and miniaturized with astonishing sophistication and success. Thus, the next big milestone will be a comprehensive catalog of BioBricks™, along with standard operating procedures guiding the computer-aided design of small biological systems, which will become increasingly bigger and more complicated. When assembling these new systems, it will be a good idea to peek into nature's bag of tricks. After all, nature had more than 3 billion years to try out designs, and if a specific task is always solved in one particular manner, there is probably a good reason for that. Even if we do not completely understand this reason, it might be good to use the same solution strategy for our synthetic organisms, at least as the first default.

Experience with both the metabolic engineering approach and the computer-aided BioBrick™ strategies of synthetic biology may even help us to create cell-free biological production systems. If this sounds like a contradiction in terms, some visionaries don't think so. The core idea is to construct a fermenter with appropriate nourishment and with selected DNA, RNA, and amino acids, as well as ribosomes, enzymes, and other molecules of importance, and possibly to add some protein or membrane structures. Ideally, this witches' brew would function like a cell, but without the need to keep anything truly alive.

The specter of quasi-living production systems or, even more so, the generation of novel organisms is obviously not without worries. In fact, if Dolly the Sheep, CopyCat and Frankenfoods don't make you at least a little bit uneasy, you are a tough soul indeed. What if we figure out how to clone people? What if Frankenfoods give us cancer? What do we do if some of our new creations take over farms and become unstoppable? Will the successes be worth the possible consequences? There is no question that science must be prudent and agree to discuss codes of ethics with elected officials and the public. After all, we are all in the same boat. However, what we are doing today and tomorrow with synthetic biology is really just a sophisticated continuation of what we have been doing in the past. Sure, we have made mistakes and we have corrected many of them. We will certainly make mistakes again, and we will hopefully have the means of correcting them. The stakes are high as we are redesigning perfection, but that's exactly what evolution has been doing for the last 3.5 billion years.

19

Let's meet in the agorá!

The agorá was the common gathering place and market of cities in ancient Greece. It was the center of activities, and nothing exhibited the pulse of the city better than the hustle and bustle in the agorá. If a philosopher felt compelled to share his wisdom with the world, if a vote were to be taken, or if one just needed to buy goods, the agorá was the place to be.

Combined with the all-important ancient concept of the division of labor, a common center for trading goods and thoughts is arguably one of the oldest and persistent creations of humankind. If I grow grain and my neighbor has sheep, it is quite natural to meet and barter. Another neighbor may trade wood or bricks, fine lace or expensive jewelry, and it soon makes a lot of sense to centralize the trading activities in a common market. While bartering is the most natural means of exchanging goods, the daily haggling soon becomes repetitive and tiring, and frequent traders develop guidelines and conversion factors, and eventually create a widely accepted currency.

Systems biology is the modern agorá in the intersection of many disciplines, most notably biology, chemistry, physics, math, computer science, and engineering. Scientists from different backgrounds meet because they see great value in solving "grand challenge" problems whose solutions are out of their own reach. It takes many bright minds and just the right confluence of knowledge and methods to address questions of cancer or the growing threat of microorganisms that are resistant to our best drugs. No single person or even discipline is capable of developing new agricultural products that can feed expanding populations, yet grow in impoverished soils under both drought and flood conditions. So the experts meet to discuss. Biologists pose complex problems and devise and execute experiments that shed light on the problem, engineers offer to create machines for collecting data more efficiently, mathematicians and physicists provide rigorous ways to formalize and address the problems with models that are anchored in a solid theoretical foundation, and computer scientists offer fast techniques for analyzing data and models. The practitioners of each discipline know that they alone cannot manage the entire spectrum of activities that are needed to solve the complex problems that arise time and again when we deal with biomedical and environmental systems. So, practitioners of the different disciplines come to the agorá to seek help and offer advice.

Anthropologists and ethnographers have studied markets for a long time. Particularly intriguing to them have been the challenges and trends that naturally evolve when the trading partners do not speak the same language. Imagine a Chinese trader traveling along the Silk Road and meeting a Turkish trader in Tashkent. Both may have goods that the other one likes. However, they are initially suspicious of each other and do not know each other's language, especially when it comes to trading objects that are unknown in their home countries. The desire to trade, and to make a good profit back home, mind you, motivates both sides to communicate, maybe by pointing and with hand gestures, or maybe with the aid of a translator. Eventually they will come to an "understanding" that results from a mixture of words from their two native languages and possibly new words not previously contained in the buyer's language. For instance, we still refer to cups and plates made from fine ceramic materials as "china" or "porcelain." During the Tang dynasty (618–906), fine porcelain from China was brought to the Islamic world, and a few centuries later to Europe. As this fine material had been unknown before, a new term needed to be introduced into the languages of the importing countries, and "china" is still the term used in many English-speaking countries today. Other countries still use "porcelain," which derives from the Italian word for a sea shell with a translucency similar to the newly imported material. As yet another alternative, a new term was derived not from an association with its origin or the resemblance to a known item, but adapted directly from the word used in its country of origin. An example from the Silk Road is "silk" itself, which resembles the Chinese word for silk: *si*.

Within their studies of trading places, anthropologists have described in detail how the original gibberish among foreign traders changes and matures over time. In the first phase of evolution, the traders use Pidgin, allegedly the Chinese pronunciation of "business." No single Pidgin language exists, because its emergence depends on trading partners, whether they bartered along the Silk Road or on a Caribbean island. A Pidgin language is a relatively simple concoction of words from different languages and new terms. It is also very malleable as new traders eventually join the discussion. If the trade is successful and becomes a recurring event, the Pidgin matures and solidifies. Certain terms become accepted and solidify as definitions of certain goods, values or situations, and a much more refined and tight "Creole" language emerges that still contains many foreign words, which, however, are seamlessly incorporated into the vernacular. A common language for the trading community has arisen.

If systems biology is a trade zone, does it have its own language? Is a language evolving? There is no doubt that English has established itself as the current language of international commerce and science, and the vast majority of learned scientific articles are published in English. For superficial discussions, even if they address scientific issues, English is quite sufficient. Nonetheless, the common vernacular becomes lacking or too vague for crisp scientific discourse. Laypersons cannot even really discuss two related insect species, because they look essentially the same to them, and scientists abandon English and resort to the strict Latin taxonomy, which the Swedish naturalist Carl Linnaeus introduced almost 300 years ago. Any common language has two problems when it comes to its use in the sciences or other disciplines that require exact definitions. First, words simply do not exist for some new items or only experts are familiar with them and their meaning. Second, many words have gained different meanings among various groups within the population. The word bug is an example. Even without its interpretation as an error in a computer code, "bugs" often seem to be synonymous with any small life forms that crawl or swarm in places where they bother us.

Due to the great diversity of its parent disciplines, systems biology is an extreme case. Math has always had the reputation of being "Greek" to outsiders, but it seems fair to say that even an extremely well-educated layperson will find understanding the abstract of a modern article on molecular biology problematic. Physics, chemistry, and engineering all have their own jargon, and listening to a computer scientist talk about new hardware specifications does not seem much different to tuning in to a radio station from a foreign land.

Do we need a common language for systems biology? Similar to the case of the Silk Road, one cannot expect that the practitioner of one discipline should have to master the languages of other disciplines in their entirety. Besides, if I go to the store to buy a digital camera, I do not have to be an expert in electronics who understands all technical details. At least to some degree it is sufficient to depend on a knowledgeable store clerk, if he is to be found, who can translate the jargon into wording that I understand.

It is clear that the involvement of a translator simplifies trade but also automatically puts both traders at a potential disadvantage. Does she use enthusiastic terms to describe what I am offering or does she sound reluctant and uninterested? How the translator selects her verbiage may obviously bias the value of the product I intend to trade. Clearly, basic knowledge of each trading partner's language would be valuable. The same situation occurs in systems biology. Every mathematical model must make assumptions, simplifications, and abstractions, and it behooves the biologist to develop a feel for what these assumptions entail. Similarly, biological data in a spreadsheet represent only those features of an experiment that are the clearest to report. Even if there is a modern push to collect metadata, which contain additional information surrounding the core data, such as details of the experimental conditions, specific strains of the analyzed organisms, machine settings, and other possible confounders, a biologist almost always knows much more about the data and their context, as well as limitations of the experimental methods used in a study, than what is documented. A mathematician or computer scientist is seldom even aware of this hidden information, let alone the biologist's gut feeling about the experiment, which is often spot on and could be very valuable for a deeper understanding of the biological issue at hand.

Because systems biology is a relatively new field, one should expect the rise and evolution of a Pidgin language within the near future. Two terms that systems biology has already requisitioned from philosophy are emergence and ontology. These terms had been used in biology very rarely, if at all, before research progress in genomics and systems biology demonstrated a need for them several decades ago. Many biologists and mathematicians probably do not have a good impression of what these terms really mean, but they have become rather common Pidgin lingo in the agorá of systems biology.

Eventually the slowly arising "systems biology Pidgin" will become a "systems biology Creole." Important components of this language will be new terms, but also ways of describing processes. An example of the latter is the need for strong correspondence rules, which form a bridge between biological reality and the realm of mathematical and computational analysis. These rules work in two directions. In the first, they take a biological phenomenon and translate it into the formal, mathematical or computational structure of a model. They define the meaning of the variables, parameters, and other settings of the model, document what they correspond to in the biological realm, what assumptions and simplifications were made in the process, and what is included or excluded from the model. In the opposite direction, the correspondence rules explain to the biologist what the details of the computational and mathematical results mean. They may translate results regarding eigenvalues into statements of system robustness and internal differences in time scales. They may interpret sensitivity values as classifiers that distinguish parameters that should be measured as accurately as possible, and those that are not very influential and therefore do not require highly accurate experimental determination.

The future "systems biology Creole" will also be instrumental for the formulation of standard operating procedures and canons of good practice. At present, the young field permits great flexibility and variability in what is researched and how experiments and analyses are performed. This flexibility is exciting and also necessary, because the burning tasks of the field are only vaguely established and strategies toward their solutions are still ill-characterized. Thus, systems biology will evolve its own language in tandem with more strictly defined procedures and a general agreement on what a practitioner of the field should know. The language will contain many words and grammatical structures from the parent disciplines but also utilize terms from other fields and develop a novel vocabulary that does not yet exist.

Historically, biologists have shied away from math, and typical current biology textbooks are essentially void of equations. Similarly, just a few decades ago "applied mathematics" was a taboo topic among "real" mathematicians, who devoted their efforts to such pure topics as algebraic topology and abstract logic. As a consequence, biology and math remained almost disjointed, with the exception of the work of a small number of pariahs interested in biomathematics. Times have changed dramatically, and many biologists and mathematicians are now eager to collaborate on some of the exciting projects that have become feasible. What is needed now, and in the next decades, is an excellent education of the next generation of students. Research in systems biology is driven by the need to formulate and solve complex, domain-crossing problems, and any educational program must explicitly aim at producing transdisciplinary, integrative thinkers. The education may be traditional or emphasize problem solving and on-demand learning. It must make optimal use of the available web resources and train students to distinguish solid information from the enormous amount of noise and falsehoods, especially on the Internet. Importantly, future education in systems biology must be multi-dialectal. It must train individuals to develop cognitive flexibility that allows them to reach across traditional scientific boundaries and address societal problems requiring novel, creative, and multi-disciplinary solutions. Because the body of knowledge in the sciences grows so rapidly and access to knowledge is becoming so easy, yet it is even easier to become overwhelmed by the entanglement of information and misinformation, the training must create a solid foundation for ongoing future learning of biological, mathematical, and computational concepts.

Even with the best training, systems biologists will not be able to master all aspects of the growing body of knowledge in the parent disciplines. Sure, future systems biologists will be expected to perform many sophisticated investigations on their own, but they will also have to rely on specialists in the biological domains of their interest and on experts in pertinent fields of math and computation. The systems biologists of this new breed will be at once independent and interdependent. They will be traders, translators, and matchmakers in the agorá, where experimental and computational scientists come to meet and discuss how to solve the "grand challenge" problems in biology, medicine, food production, and the sustainable stewardship of the environment.

20

Dessert

"Are we there yet? When are we gonna be there?" Who does not recall the little voices of backseat drivers a couple of hours into a 300-mile trip? We smile, and somewhat exhausted may offer a platitude like "We'll get there when we get there," which may grant us a short reprieve, although we know full well that it will only confuse a young, inquisitive mind.

Systems biology is in quite a similar situation. The young endeavor has been receiving a lot of attention, and excitement about its potential has led to widespread enthusiasm and support. These are very positive developments indeed, but we must realize that they also come with a hidden danger: the public, the funding agencies, and even colleagues from more traditional biological fields are not infinitely patient, and systems biology must expect that the initial hype and sometimes exaggerated promises may eventually turn into demands for major tangible successes. We have invested a lot; what have we gotten in return? Why has systems biology not yet cured cancer? Except for a very few isolated cases, personalized medicine is still a distant fantasy; when will we enjoy its fruits? Systems and synthetic biology still look like academic, if not totally esoteric, pursuits; what have they produced that we could not have achieved with traditional means?

There is no doubt: systems biology is not "there yet." Maybe worse, it is entirely unpredictable when we are "gonna be there." And in contrast to small children, the public and the national funding agencies can seldom be mollified with statements such as "Trust us, we'll get there" … with the occasional addition of "eventually" under our breath. Yes, we are a long way away from "there," but we must not let doubt distract us from our noble quest. Systems biology is still young and simply has no choice but to stay the course and march onward, while asking for patience and carefully documenting our initially modest successes along the way. After all, we systems biologists are convinced to the core of our being that there is no alternative to dedicating our most fervent efforts to advancing this exciting field of endeavor, and we can indeed point to some early successes. Sure, we could instead continue to generate hundreds or thousands of –omics datasets, but what good would that do if we did not have the tools of analysis to make effective use of them? As we discussed throughout this book, we can hardly trust our intuition or merely think very hard when tasked with integrating the huge amounts of data that are producible with modern biological methods, let alone with future technologies. We need machine learning to find patterns and associations. We need sophisticated statistics to infer from experimental data the structures of systems and the processes that govern them. We need mathematical and computational methods that give us insights into static networks and adaptive and highly regulated dynamical systems. And, of course, we need comprehensive experimental methods that are capable of quantifying more variables across more systems and more scales. With what we know today, it is hard to imagine a future of biology and medicine without the tools of systems biology.

So we have embarked on the journey with what future generations will consider baby steps, and we know that the journey will be long. But where exactly is "there"? Where do we want to be in 20 years' time? Do we have clear goals? Is there a Holy Grail of systems biology? Do we even have reliable guideposts along the way?

This being science, the paths and destinations are driven by individualists, many of whom just know that they know best, and there is therefore of course no one common answer. Nonetheless, the typically stated goals fall into two categories. One calls for very big models and the other for rather small ones. At the same time, most systems biologists will probably agree that both pursuits start in the middle, with moderately sized *mesoscopic* models that we can tackle with our current repertoire of methods and techniques.

The proponents of the first category call for whole-cell models or even more complicated models, for instance, of entire organisms or complex diseases. Clearly, computational models that can explain a disease and suggest specific, reliable, and maybe even personalized treatments require thousands of variables and well-characterized processes. Indeed, some companies today are already attempting to collect all available pieces of information pertaining to a disease and to integrate them into computational models that pharmaceutical companies may use to screen for possible drug targets and specific treatments. Of similar importance, reliable whole-cell models could possibly answer innumerable, fundamental questions that have nothing to do with diseases, but with how organisms function and respond to changing environments. And again, some groups are already claiming to have developed whole-cell models, even if only for specific aspects or for relatively simple cells.

At the other end of the spectrum, proponents of small models argue that we truly need to understand the fundamental principles that govern the functionality of biological systems. As discussed in Chapter 4, the rationale for pursuing issues of design and operation is that we should probably not manipulate, let alone create, organisms without a deep understanding of why biological systems are organized in specific ways and not in an alternative fashion. After all, nature had a lot of time to try out different options and, as Thomas Edison famously said about his own trials and errors, nature did not fail 10,000 times with early designs that were eventually replaced by evolution, but instead figured out thousands of ways to improve the best design that was available at the time.

Discovering the rationale for specific design and operating strategies is only the first step in this line of research. The ultimate goal is the discovery of general laws or, better, theories for certain aspects of biology. Searching for theories might sound like the paradigm of an academic ivory-tower activity, but the establishment of a theory has far-reaching implications. As Kurt Lewin's famous saying goes, there is nothing more practical than a good theory. Indeed, if we look at our role model, physics, we easily catch a glimpse of how powerful a good theory can be: because physical theory describes and quantifies how mechanical and electrical forces work, we are able to send spacecraft to Mars or Pluto, and they actually go where we intend them to go, even though we had never tried these tasks before. They follow the trajectories predicted by the theories of physics and mathematics. Just imagine this type of predictive power in biology and medicine!

At present, biology does not have much to offer in terms of theories or laws. Nonetheless, remember the genetic code as one already established example of a "law." It poignantly indicates the practical importance and applicability of the generality of such laws. As we discussed elsewhere, the "codon law" associates triplets of DNA nucleotides with specific amino acids. As a consequence, we can reliably infer from a piece of coding DNA what the corresponding protein will look like. If we did not have this law, we would have to decode the DNA from every species, or even every DNA strand, anew! Instead, relying on the validity of the law we are able to compare sequences from different organisms quite easily and, at least to some degree, can infer the role of a gene in a "new" organism from what we know about other organisms.

The two long-term goals thus discussed require systems biology to move in two opposite directions at once: toward very detailed, large models; and toward small models that focus on the essential features of a natural design. Right now, sufficiently detailed models of organisms and diseases seem overwhelmingly difficult to achieve, and the search for design motifs is limited to rather small systems that have been stripped of most of their non-essential details. Thus, casual consideration suggests that the two pursuits are independent and overlap only minimally. However, at a deeper level, they actually shine light on two sides of the same coin, because truly understanding the design and operational strategies of nature will eventually help us tremendously with the analysis of large, complex systems, and hopefully prevent us from making grave mistakes on the road toward manipulated or newly assembled organisms. Also, both pursuits explicitly or implicitly attempt to determine what the most appropriate models and system representations are. When setting up a model, what is the optimal level of complexity? How do we find the best balance between accuracy and feasibility, between capturing the most details and gaining deeper insights into the inner workings of the system in question? Might it be possible to design truly correct models?

For instance, when designing a model of the heart, where do we learn the most? From a physical or mechanical model based on the principles of a peristaltic pump? From a model focusing on electrical, chemical or mechanistic aspects? From a model describing oscillations and derailments of these oscillations that lead to fibrillations? From a model that somehow merges all these aspects into a very complex supermodel?

We discussed that the most appropriate model of a phenomenon or system depends on its purpose and on the questions that we would like to answer. Is it imaginable that a sophisticated, futuristic supermodel may potentially be able to answer all such questions? Or are there unknown, immutable principles mandating different types of model for questions of optimal design and the search for maximal realism? At this point, we just don't know.

Whatever the answers to all these questions might be, it will not come as a surprise that we will not reach either goal anytime soon, and many a systems biologist pursuing these goals today may only reach Mount Pisgah without ever reaching the Promised Land and enjoying the rewards of all the efforts along the way.

The exciting adventure of systems biology has just begun, and as always in science, the path is seldom straight and often quite foggy beyond the immediate next steps. But instead of worrying about the unknown, we shall celebrate the journey itself, agreeing with the eighteenth-century German poet Johann Wolfgang von Goethe, who traveled through Europe and opined that, "you certainly would not travel just to get there."[1] So let's move onward, enjoy to the fullest the twists and turns that lie ahead, offer bidialectical education to the next generation of scientists, and always encourage our students and peers to experience the living world as a complex system of complex systems. Let us savor the sweet taste of victory over obstacles and challenges in our travels and celebrate the tasks accomplished.

1 Translation my own.

Gentle jargon

Abstraction	The intentional omission or simplification of details of a phenomenon, which allows the design of a manageable conceptual, mathematical or computational model. A representation that focuses on details relevant to the question of interest.
Activator	A molecule or entity that increases the capacity of a process.
Ad hoc model	A mathematical model that is constructed for a specific purpose and without adhering to prescribed model design rules, as opposed to a canonical model.
Adaptive system	A computational or natural system that adjusts its own internal settings to improve its tolerance toward changes in inputs or in its environment.
Affinity	An attractive force between entities. In the case of atoms or molecules, affinity may lead to a chemical reaction.
Agent-based modeling	A simulation method for discrete models, in which each component is an agent whose actions are determined by specific rules. Agents may be humans, animals, cells, molecules or even single atoms.
Algorithm	A sequence of operations or commands constituting a numerical procedure or computer code.
Alkaline	A chemical or solution that is basic, with a pH greater than 7.
Alternative splicing	A mechanism allowing peptides that were transcribed from the same gene and translated from the same mRNA to be combined into different proteins.

Amino acid	The molecular building blocks of proteins. While about 500 amino acids are known, only 22 are used in proteins, and only 20 are directly represented by the genetic code.
Antagonism	A situation where two components act against each other or diminish each other's effect. Antonym of synergism.
Antibody	One of many key proteins of the immune system that sense and neutralize foreign molecules and organisms in the blood. Antibodies are very specific in their binding to non-self entities. Also called an immunoglobulin.
Apoptosis	The technical term for programmed cell death. A complex process in which a cell is genetically programmed to die under certain conditions.
Approximation	The replacement of a (complicated) function with another (simpler) function that retains some, but not all, important features. Usually, the approximation and the approximated function have the same value and the same slope(s) at one point that can be selected as most important for representing a particular situation.
Artificial neural network (ANN)	A type of machine learning algorithm used to discover patterns in complex datasets. As an example, given a large enough dataset, an ANN may be trained to identify genes associated with a disease.
Astrocyte	By far the most abundant cell type in the brain, performing a variety of support functions for neurons and the blood–brain barrier.
Autoimmune disease	Abnormal condition in which the body is attacked by its own immune system. Examples are lupus and Crohn's disease.
Axon	A single, long appendage of a nerve cell that transmits signals away from the cell body. See also, *Dendrite*.
Bacteriophage	A virus that infects bacteria.

Behavior (of a system)	The qualitative or quantitative collection of the dynamics of a system, often in response to a stimulus or perturbation.
Bidialectical	Able to converse in two languages or substantially different dialects.
Bifurcation (in a dynamical system)	A threshold value of a parameter at which the behavior of a system changes qualitatively. In the most typical case, the system starts to oscillate once the value of the parameter is increased above this threshold.
BigData	New buzzword describing collectively the huge numbers of data that can be obtained with modern experimentation, social media or other acquisition methods. BigData are so complex that they require machine learning methods for their analysis.
BioBrick™	An artificial module, made from actual biological components, which can be connected to other modules in order to form a more complex, functional system.
Biofilm	A thin layer, for instance in wet pipes or on wet surfaces, consisting of microbes that often belong to hundreds if not thousands of different species. A biofilm is a special case of metapopulation.
Bioinformatics	A scientific field of study using methods from computer science to organize, analyze, and interpret large-scale biomedical datasets, in particular from –omics studies.
Biomarker	A biological entity, such as a protein, which, when sufficiently outside a "normal" range, is associated with a disease or some other specific situation. The alteration of the biomarker may be the cause of a disease or just a symptom.
Blue-sky catastrophe	An example of deterministic chaos, consisting of a single differential equation. The solution apparently oscillates regularly but, without any perceivable hints or warnings, can switch to a different range of oscillation and back.

B-lymphocyte (B-cell)	In most cases, a short-lived cell type of the adaptive immune system that produces antibodies in response to antigens. However, some of these cells mature into long-lived memory B-cells that remember antigens they have encountered before.
Bradykinin	A signaling protein causing blood vessels to enlarge.
Canonical model	In contrast to an *ad hoc* model, this mathematical modeling format follows strict rules of model design, diagnostics, and analysis. These rules may seem limiting, but are often theoretically and practically advantageous, as they permit the use of customized analytical tools. Maybe surprisingly, canonical models are extremely rich in their range of possible behaviors.
Causality analysis	In contrast to an analysis of correlations, causality analysis is a statistical assessment of the likelihood that one component of a system effects – or is responsible for – changes in other components.
Cell cycle	A well-coordinated sequence of steps ultimately leading to the dividing of a cell into two daughter cells. It involves the duplication of important contents of the cell, in particular its DNA, as well as their assignment to typically two daughter cells that result from the cell division at the end of the cycle.
Cell division	The end of a cell cycle and a key process for the growth of a cell population. In most cases, the "mother" cell splits into two daughter cells. In many cases, these are quite similar. However, in the process of differentiation, one daughter cell is typically similar to the mother, while the other daughter is substantially different. In some species, such as baker's yeast, cell division essentially retains the mother cell, while a smaller daughter cell forms as a "bud."

Central dogma (of molecular biology)	The concept that DNA is transcribed into mRNA and mRNA is translated into proteins. Some of these proteins catalyze biochemical reactions that create or change metabolites. While the central dogma is basically correct, the process is in reality not unidirectional, but much more complicated. For instance, different RNAs, proteins, and metabolites can affect or even control transcription, and some RNAs can serve as enzymes. Also, some viruses contain only RNA, which in the host is "reverse-transcribed" into DNA.
Cerebellum	Part of the brain in the back of the head responsible for the control of muscle activity.
Cerebrum	Most of the brain, located in the front of the head and consisting of the left and right hemispheres. It is responsible for interpreting signals from the outside and for voluntary responses.
Chaos (deterministic)	Erratically oscillatory behavior of a nonlinear system that is unpredictable in the long term, even though the system is entirely deterministic, all parameters and settings are numerically specified, and the erratic behavior can be explored through simulations.
Chaperone (protein)	One of several proteins that facilitate the formation or disassembly of large molecular structures.
Chromosome	Carrier of genetic information within the nucleus of animal and plant cells. Each chromosome consists of long stretches of DNA that are coiled around histone proteins, which allow tight packaging and also affect transcription.
Circadian rhythm	A collection of oscillatory changes in physiology with a frequency of about 24 hours.

Cluster analysis	A tool of statistics and machine learning for categorizing objects within a large set into groups with similar features.
Codon	A set of three nucleotides on a DNA strand that codes for an amino acid when the triplet is transcribed and translated.
Complement system	A complex group of molecules within the immune system that assists with the removal of foreign invaders.
Complexity	A combination of features of a nonlinear system that makes predictions regarding its behavior difficult. Often complexity is the result of many interwoven regulatory processes.
Computed tomography (CT) scan	A sophisticated computational method for converting X-rays, taken from many sides, into a series of cross-sectional images of parts of a body.
Conformation	The three-dimensional spatial arrangement of atoms in a protein or other large molecule.
Constraint (optimization)	A restriction within an optimization task. In many cases, a constraint is a numerical limit that a variable is not allowed to exceed.
Correlation	A mutual relation or interdependency between two objects, which however does not necessarily imply causality.
Correspondence rule	A defined association between something in the real world and a component of a model. Correspondence rules include the definition of variables and help with the interpretation of model results.
Cortex	The outer layer (grey matter) of the brain.
Creole	A new, emerging language resulting from the mixing of other languages.
Cross-talk	Exchange of information between two systems. The term is often used in the context of signaling cascades.

Cytokine	A compound within a class of proteins that is excreted by tissue or some white blood cells, signaling an inflammation or infection. While these signaling proteins are of extreme importance for the immune response, their overproduction can lead to a fatal "cytokine storm."
Cytosol	The water-soluble inside of a cell, in which organelles like the nucleus and mitochondria are suspended.
Data mining	Application of statistical machine learning techniques for identifying patterns or extracting information hidden in large datasets.
Dendrite	A heavily branched extension of a nerve cell that receives signals from other nerve cells
Design (of a biological system)	and transmits them to the cell body. See also, *Axon*. The natural or synthetic, often optimized, blueprint or construction of a system.
Design principle	A feature of a system that is observed more often than one might expect, usually due to its functional superiority. Similar in usage to a motif.
Deterministic (model)	Not involving or allowing for any randomness. Antonym of stochastic.
Difference equation	Core building block of a discrete system, the behavior of which is determined by a mathematical function of the state of the same system at one or several earlier time points.
Differential equation	Base unit of most dynamical systems, equating the change in a variable to all processes that increase or decrease this variable.
Differentiation (cell biology)	A change from one cell type into another, usually during an asymmetric cell division. Typically, one daughter cell is a stem cell, like its mother, whereas the other daughter cell is quite different in function, abilities, and appearance.
Discrete	Considered at distinct time points or for distinct points in space, but not in-between.

DNA microarray	Experimental tool for comparing the expression of genes in two systems. Also called a DNA chip or gene chip.
Dopamine	A neurotransmitter associated with rewards and learning, as well as movement control.
Dynamical system	A real-world system whose state or behavior may change over time; a mathematical model of such a system.
Dynamics	The time-dependent behavior of a system or model.
Edge (of a network)	A uni- or bi-directional connection between two nodes of a network, representing some type of relationship.
Eigenvalues	A set of mathematical features that are, among other properties, indicative of the stability of a system at a steady state.
Electrocardiogram	Result of a diagnostic test measuring the electrical activity in and around the heart.
Emergent property	A feature of a system that cannot be explained with any of its components alone but is the result of the interactions between two or more components.
Enzyme	A protein that renders a biochemical reaction possible. Most metabolic conversions are catalyzed by specific enzymes.
Eosinophil	A type of white blood cell, and a component of the immune system, that fights infections and foreign organisms in the blood.
Epigenetics	A sub-field of genetics that studies biological features that are not inherited through genes but passed on to children through the attachment of specific molecules to the inherited DNA and/or alterations in DNA packaging on histones. Epigenetic changes can alter the expression of nearby genes.
Epithelial–mesenchymal transition	A dramatic change from an epithelial cell, which lines the surfaces of blood vessels and bodily cavities, into a stem cell that can

turn into a variety of cell types. A mechanism underlying processes of wound healing and cancer metastasis.

Error (in modeling and statistics)
A measure of dissimilarity between a model output and (experimental) data. Also called residual.

Eukaryotic
Pertaining to a cell that contains a nucleus or to an organism made from such cells.

Evolutionary algorithm
An optimization method inspired by natural evolution and mathematically based on machine learning. The most prevalent example is a genetic algorithm.

Explanatory model
A model causally relating phenomena at a high level (e.g., physiology) to mechanisms at lower levels of organization (e.g., gene expression).

Extrapolation
Use of a model to predict system behavior outside the data range for which the model was created.

Feedback
Signal sent by a downstream component of some pathway or chain of events that affects the dynamics of an upstream component.

Feedforward
Signal sent by an upstream component of some pathway or chain of events that affects the dynamics of a downstream component.

Fibrillation
An extended sequence of erratic contractions of muscle fibers, particularly in the heart.

Fit (of a model)
Measure of closeness or similarity between a model result and data.

Fitness
In evolutionary biology, a measure of reproductive success. In optimization, a measure of quality of the representation of data by a model in comparison to other models.

Flow cytometry
An automated method for counting and sorting cells based on natural or artificially added markers on their surfaces. Also used for quantifying such markers.

Fluorescence	The ability of molecules and materials to emit visible or invisible light when stimulated with light of a shorter wavelength, such as ultraviolet light.
Flux	Amount of mass flowing through a reaction or variable pool per time unit.
Functional magnetic resonance imaging (fMRI)	A noninvasive method for measuring brain activity that is associated with changes in nearby blood flow. See also, *Magnetic resonance imaging*.
Gas chromatography	A method of analytical chemistry that separates and quantifies the amounts of target compounds in a sample.
Gene circuit	A group of genes that affect each other's expression, mostly through transcription factors.
Gene ontology (GO)	A collection of strictly defined terms for the interpretation of genes and their functions.
Genetic algorithm	A computational optimization method inspired by genetic processes and evolution. See also *Evolutionary algorithm* and *Optimization.*
Genetic code	The natural assignment of amino acids to DNA codons. Each codon consists of a triplet of nucleotides that is transcribed into an RNA nucleotide triplet and translated into a specific amino acid or into stop and start signals. Knowledge of a coding gene sequence is necessary to predict the composition of the corresponding protein. It is not quite sufficient for a complete prediction, as translated pieces of proteins may be spliced together in different ways.
Genome	All of an organism's own DNA, usually excluding the small amounts of DNA found in mitochondria.
Genome-wide association study (GWAS)	A massive analysis of genetic variants within whole genomes of large numbers of individuals with the goal of associating certain

combinations of genes with observed traits, such as body size or a disease.

Genotype
: The genome of an individual. See also, *Phenotype*.

Ghrelin
: A signaling protein that is secreted by the stomach and acts on the hypothalamus in the brain, where it creates the feeling of hunger. See also, *Leptin*.

Glutamate
: A widely used neurotransmitter involved in cognition, learning, and many other functions.

G-protein-coupled receptor
: One of many proteins, inserted in the cell membrane, that can bind specific molecules outside a cell and signal their presence via signaling cascades inside the cell to the nucleus, thereby triggering changes in gene expression or cellular function.

Graph
: A visualization or formal representation of a network through nodes (vertices, hubs) and connections (edges, arrows). An example is the route map of an airline.

Hermaphrodite
: An organism possessing both female and male sex organs.

High-throughput
: Vaguely defined term referring to very many analyses or experiments that are performed at the same time. Often such experiments are executed by a robot in an automated fashion, either in parallel or sequentially.

Histamine
: A protein of the immune system that is involved in responses to inflammation and facilitates the movement of white blood cells and fluids from capillaries into infected or inflamed tissue, sometimes leading to an allergic reaction. Histamine can also serve as a neurotransmitter.

Histone
: One of several proteins around which DNA is wrapped in eukaryotic cells.

Hodgkin–Huxley model	The first – Nobel Prize-winning – mathematical model of the electrical "firing" activity of nerve cells, formulated as an electric circuit.
Homeostasis	The normal state of a cell or organism, which is maintained through uncounted regulatory processes. A physiological steady state.
Horizontal gene transfer	The migration of DNA between different species. Among bacteria, the transfer is often mediated by viruses or small rings of DNA, called plasmids, that can be shared among microbes.
Humoral	Referring to body fluids, such as blood serum.
Hybrid vigor	Superiority of offspring that inherited genes from rather different parents.
Hypochlorite	A highly aggressive compound consisting of chlorine and oxygen, produced in the body by monocytes and neutrophils to damage foreign proteins. The active compound in household bleach.
Hypothalamus	A relatively small brain portion located at the bottom of the brain and controlling many bodily functions, including hunger and thirst, body temperature, sleep, and the circadian rhythm.
Immunodeficiency	A pathological state of reduced or entirely lacking immune function.
Immunoglobulin	A key protein type of the immune system that senses and neutralizes foreign molecules and organisms in the blood; also called an antibody.
Inducer	A molecule that turns on or upregulates the expression of a gene, often by either blocking a repressor or binding to an activator.
Input	A collective term for quantities that are fed into a system or model.
Interferon	A signaling protein of the immune system warning cells of the presence of viruses, bacteria or other foreign organisms. Interferons

can also interrupt the replication of viruses and thereby protect the host against them.

Interleukin
A signaling protein of the immune system that enhances the production of T-cells and B-cells, as well as cells of the blood-building system.

Ion channel
A pore-shaped protein embedded in the cell membrane that allows the movement of ions like sodium, calcium, and potassium. This flow of ions, in turn, can be a mechanism of electrical signaling, for instance in the heart.

Iteration
A sequence of repeatedly executed steps within a computer algorithm.

Kinase
An enzyme that attaches a phosphate group to a molecule. See also, *Phosphatase*. Kinases and phosphatases are crucial components of signaling cascades.

Kinetics
(in biochemistry)
The field of study investigating the speed and regulation of biochemical reactions.

Laser capture
microdissection
A relatively recent method of isolating individual cells from a tissue. The isolation is performed under a microscope with a laser.

Learning (supervised)
In this type of machine learning, the predictions of the computer algorithm gradually improve based on comparisons with known, correct and incorrect answers.

Learning (unsupervised)
In this type of machine learning, correct and incorrect answers are not known, and the task of the computer algorithm is to detect so-far unrecognized patterns or associations within large datasets.

Leptin
A signaling protein, secreted by fat cells, that counteracts the feeling of hunger. See also, *Ghrelin*.

Leukotrienes
Signaling molecules that are not proteins, but use lipids for signal transduction in cases of inflammation.

Limit cycle	A sustained oscillation with the same amplitude and frequency. If slightly perturbed, the oscillation recovers and reenters the earlier oscillatory pattern.
Linear	Straight or representable using a sum. Linear dynamic systems usually express the change in each variable as a sum of processes. These systems have numerous mathematical advantages and can be analyzed with a much richer repertoire of methods than nonlinear dynamic systems.
Linear algebra	A large, extremely well-understood branch of mathematics using vectors and matrices to solve problems represented by linear equations.
Lymphocyte	A white blood cell of the immune system in the form of a natural killer cell, a B-cell or a T-cell.
Lysozyme	An enzyme that can damage the cell walls of bacteria and is found in fluids like tears and saliva, as well as breast milk.
Machine learning	A collection of methods using large-scale computing, artificial intelligence, and statistics to extract information or characterize patterns hidden in large datasets. A key tool of BigData analysis.
Macrophage	A large white blood cell that can devour bacteria or other foreign organisms, cell debris, and other small particles.
Magnetic resonance imaging (MRI)	A noninvasive method using a magnetic field to create images of organs and other features inside the body.
Major Histocompatibility Complex	A huge collection of different molecules that are attached to the surfaces of cells. The molecules distinguish self from non-self and are therefore critical for proper immune responses.

MALDI-TOF	Matrix-Assisted Laser Desorption/Ionization – Time of Flight. A method of mass spectrometry that first attaches a biological sample to some standardized material (the matrix), and then uses a laser to separate (desorb) molecules from the matrix, thereby converting them into ions, which are shot through an electromagnetic field toward a detector. The arrival time is a measure of the mass of the molecule.
Mass-action (kinetics)	The first reported and still widely used, over 150 years old, mathematical formulation of a chemical reaction.
Mass cytometry	A mass spectrometric method that permits measurements of features, such as the phosphorylation state of specific proteins, in single cells.
Mass spectrometry (MS)	Experimental separation technique for molecules of the same type, such as proteins or metabolites. Modern methods of MS can distinguish thousands of different species of molecule from a small amount of a biological sample.
Mesoscopic	Of an intermediate scale; between micro- and macroscopic.
Metabolic burden	A limitation in metabolic engineering caused by the fact that foreign DNA introduced into a (micro-) organism requires ribosomes, proteins, and other molecules to be functional. These additional molecules incur a "cost" to the organism, as they must be synthesized and maintained.
Metabolic engineering	A field of study within chemical engineering and bioengineering with the goal of manipulating (micro-) organisms or higher cells into producing valuable compounds, such as amino acids, precursors for prescription drugs, or alcohol.

Metabolic pathway	A sequence of biochemical reactions, usually catalyzed by enzymes. Collections of such pathways form metabolic networks and, when including regulation, metabolic systems.
Metabolomics	The field of study addressing large collections of metabolites simultaneously.
Metadata	A comprehensive description of datasets, including details of the experiments, sources of the materials used, and other features that could potentially lead to undesired uncertainty in the data and need to be considered during data analysis and if the experiment is replicated.
Metagenomics	An extension of genomics toward meta-populations. Metagenomics is often more interested in which genes are present in a metapopulation than to which species a particular gene belongs.
Metapopulation	A large assembly of (micro-) organisms from many different species, found in soils, lakes, and oceans, but also as one of many microbiomes in bodily spaces such as the oral cavity, the gut, and the vagina.
Michaelis-Menten mechanism (rate law)	A very widely used concept and mathematical representation of a biochemical reaction that is catalyzed by an enzyme.
Microarray (technology)	An experimental laboratory on a glass or silicon chip for assessing the abundance or expression of DNA, microRNAs, proteins, cells, antibodies or other entities in a cell or sample. The best known type is the DNA or gene chip.
Microbiome	A collection of many populations of microbial species living together in or on the human body or another higher organism. A microbial metapopulation.
Microfluidics	A multidisciplinary branch of science that uses minute channels and devices on a silicon

platform to execute biological or other experiments at the micrometer scale.

Micro-RNA
One of thousands of short RNA strands known to play specific regulatory roles, for instance by silencing a gene.

Mitochondrion
An organelle within a cell of a higher organism that is responsible for energy production.

Model
1. In biology, one of numerous particular species that serve as representatives for a wide range of organisms, such as a "mouse model." Other prevalent models are the bacterium *E. coli*, baker's yeast, the worm *Caenorhabditis elegans*, and the fruit fly *Drosophila melanogaster*. In addition, cellular model systems are used, for instance, to mimic a disease process. 2. In systems analysis, a conceptual, mathematical or computational formalization of a phenomenon or system.

Modeling
(in systems biology)
The act of designing, diagnosing, analyzing, and interpreting a conceptual, mathematical or computational model of a biological phenomenon.

Model reduction
The attempt to capture the essence of a system in a model with a lower number of variables than originally used.

Module
A clearly defined, by itself, functional sub-model within a larger model or system.

Molecular modeling
A branch of physico-chemical theory and computational modeling that attempts to understand biochemical processes at the scale of individual molecules.

Monocyte
A white blood cell of the immune system that can quickly move to a site of infection and turn into a macrophage.

Monte Carlo simulation
A type of systems analysis that uses thousands of solutions of the same model, obtained with different, randomly sampled

parameter settings, in order to establish possible, likely, and worst-case behaviors of a complex system.

Morphogen
A signaling molecule affecting the spatial arrangement of cells during development.

Motif
A surprisingly often observed structural feature within a biological system. A design principle.

Motor nerve
A nerve sending signals from the brain to a muscle.

Mucus
A substance covering many surfaces exposed to the environment, such as in the oral cavity. Mucosal layers contain uncounted beneficial bacteria and viruses, along with enzymes and other compounds that block microorganisms from invading.

Mutual information
A measure of how much knowledge about one variable is conveyed by knowledge of another variable.

Natural killer cell
A type of lymphocyte responding quickly to cells that are stressed, for instance by a virus infection.

Network
An assembly of connected items (nodes, variables, pools, boxes), often represented as a graph. The connections (edges, arrows) may point in one direction or may be bidirectional. As an example, a gene regulatory network contains genes as nodes and the effects of each gene on the expression of others as edges.

Neuron
A nerve cell, consisting of a cell body, a long axon, and dendrites that are branched many times.

Neurotransmitter
One of a number of chemicals with which neurons transmit signals to each other. Examples include dopamine, glutamate, noradrenaline, and serotonin.

Neutrophil	Most abundant white blood cell and first responder in cases of infection or inflammation.
Niche (ecology)	A space characterized by specific physical, chemical, and biological features that encourage some species to flourish and others to disappear.
Nitric oxide	A small signaling molecule consisting of nitrogen and oxygen.
NMR (nuclear magnetic resonance spectroscopy)	A method, based on the movement of protons and neutrons within atoms, for assessing the structure of proteins and for measuring metabolites within living cells.
Node (of a network)	Also called a vertex. An entity, as opposed to a relationship, within a network or graph, which may be connected through edges to many, a few, one or no other nodes (vertices). As an example, each metabolite is a node in a metabolic network and connected through biochemical reactions.
Noise	Usually unwanted and often unexplained deviations in data from expectations. See also, *Error* and *Residual*.
Nonlinear	Not linear, curved; not representable exclusively with sums. Nonlinearity is a basic ingredient of complexity and of the phenomenon of synergism as it violates the law of superposition.
Nonlinear system	A natural or mathematical system or model containing or exhibiting nonlinear features.
Nucleotide	One of five fundamental molecular building blocks of DNA and RNA. All genes are composed of DNA nucleotides containing adenine (A), cytosine (C), guanine (G), and thymine (T). In RNA, uracil (U) replaces thymine.
Nucleus	Key organelle of higher cells containing most of the genetic material.

Oculomotor nerve	The nerve controlling muscles responsible for most movements of the eye.
–ome, –omics	Modern buzz-suffixes referring to the totality of molecules or other items of a similar type. The genome consists of all genes, and the interactome includes all interactions of a particular type, within a cell or organism.
Ontology	Strictly defined, often computer-interpretable, vocabulary of terms associated with some sub-field of knowledge. Probably the best established ontology in biology is the gene ontology (GO).
Operating principle	An often observed, advantageous strategy, used by nature or a modeler, to solve a recurring task. The dynamic analog of a design principle or motif.
Operon	Arrangement of several genes next to each other within the DNA of a bacterium. The same promoters and repressors control the expression of all genes within the operon.
Optimization	Determination of the "best" set of values among many possible values. A prominent case is parameter estimation. In simple cases, optimization is possible with methods of calculus, but practical optimization tasks are almost always performed with computer algorithms, of which there are many varieties.
Ordinary differential equation	A key format for dynamic systems in biology. In this type of equation, the change in each system variable is expressed as the sum of all processes that directly affect this variable.
Organelle	One of many compartments within a cell of a higher organism that are enclosed by lipid membranes. Examples include the nucleus and mitochondria.
Oscillation	System behavior in which a variable swings back and forth between low and high values. Depending on the type of oscillation, the

	difference between the low and high values may remain the same, increase or decrease. See also, *Limit cycle*.
Output	A collective term for quantities or behaviors that a model exhibits.
Overfitting	Apparently very good, yet undesirable fit between a model and actual data, caused by the inclusion of too many parameters in the model. While the fit for the particular data is good, which is of course desirable, the quality of the model representing or predicting other, similar data, may be quite bad.
Paradigm shift	A fundamental change in the way a problem is considered and assessed.
Parameter (of a model)	A quantity within a model whose numerical value is kept constant for a given analysis or simulation but may be given a different value for other simulations. Often, the structure of a model, such as the pattern of connections between nodes, is fixed, whereas different sets of parameter values allow changes in the strength or importance of each connection.
Parameter estimation	An important subspecialty of optimization, whose objective it is to determine values for parameters that make a model result fit given data as well as possible.
Pattern recognition	A machine learning technique for categorizing patterns or data.
Peptide	A short protein strand with fewer than about 30 amino acids.
Perforin	A protein that makes holes in the membranes of cells that are tagged for destruction. Water and ions enter through these holes and cause the cells to burst.
Peroxide	A small, aggressive molecule, consisting of two oxygen atoms, that can react with many organic molecules.

Personalized medicine	An emerging frontier in medicine, whereby treatment is customized for a specific individual, in accordance with his or her genomic, proteomic, metabolic, and physiological state.
Perturbation	A natural or artificial, often rapid, change in inputs affecting a system or model.
PET (positron emission tomography) scan	An imaging method that uses a radioactive tracer and creates three-dimensional representations of processes, such as an increased use of glucose by cancer cells, inside the body.
Phagocytosis	The process by which a cell engulfs some solid particle or bacterium.
Pharmacogenomics	The assessment of the role of genes in the response to drugs and of interactions between drugs, gene expression, and subsequent alterations in physiology.
Phenotype	A vaguely defined term describing the appearance or physiological state of an organism. The phenotype depends on the genotype, but different genotypes may lead to the same phenotype.
Phosphatase	An enzyme that removes a phosphate group from a molecule. See also, *Kinase*.
Phosphorylation	Attaching a phosphate group to a protein or some other molecules. The phosphorylated form of a protein is often active, thereby allowing the protein to catalyze a biochemical reaction or to send a signal, whereas the dephosphorylated form is often inactive.
Phylogenetic	Pertaining to the genetic similarity of different species within the context of evolution.
Pidgin	A primitive conversation language used by parties that natively speak languages that the other party does not know.
Pluripotent	The ability of a stem cell to differentiate into several, although not all, cell types of an organism.

Power-law function	The product of a constant multiplier and one or more variables, each raised to some exponent.
Prediction (of a model)	An output or internal feature of a system that is computed with a model, but has not yet been experimentally measured.
Predictive health	An emerging frontier in healthcare research that attempts to predict the future health status of an individual based on personal biomarkers and the individual's health history and current status.
Probability theory	A branch of mathematics addressing random events or processes. The foundation of statistics.
Profile (of metabolites or proteins)	A set of (metabolite or protein) concentrations at one time point or at a sequence of time points.
Prostaglandin	One of several lipids with a specific molecular structure having signaling functions, similar to hormones. Derivatives of prostaglandins can enlarge or restrict capillaries, for instance, during inflammation.
Proteasome	A large protein complex in cells responsible for the destruction and removal of old or misshaped proteins.
Proteome	Collectively all of a cell's or organism's own proteins.
Pyrogen	A molecule causing fever or local warming of a tissue, for instance during an infection or inflammation.
Quantum dot	One of many very small crystals that can be used for imaging proteins in biological experiments and for diagnostic purposes. Quantum dots have been created in many different colors.
Quorum sensing	A chemical signaling system used among microbes: if sufficiently many microbes within a population release a specific chemical

signal, the entire population exhibits a collective response. For instance, the microorganisms may aggregate into what looks like a single multicellular organism.

Random   Stochastic, unpredictable, such as the outcome of rolling dice.

Reaction   The base process of metabolic and other bio-
(in biochemistry)   chemical systems, in which a substrate molecule is converted into a product molecule, often facilitated by the action of an enzyme. The speed of a reaction may be modulated by activators and inhibitors.

Receptor   The sensing component of a signaling system, usually consisting of a protein that is embedded into a cell membrane and sticks out of the membrane to the outside and the inside of the cell. A signal is transmitted if molecules, called ligands, bind to the outside portions of receptors in sufficient numbers.

Reductionism   A widespread concept in biology proffering that the function of a system can be understood by studying, in sequence, the system's components, the components of these components, and so on, until the most basic building blocks are reached. Systems biology complements reductionism by studying the interactions of components at some level and the resulting (emerging) properties at a higher level of organization, which are the consequences of the genuine complexity of the system.

Regression   A set of widely used methods of statistics that match a model to data such that the residual error is as small as possible.

Regulon   A group of operons that are under the control of the same regulatory proteins.

Repressor   A protein that blocks the expression of a gene.

Residual (error)	The error or dissimilarity remaining after the parameters of a model have been (partially or fully) optimized to fit a dataset. This error is due to natural or experimental noise in the data and/or to a model structure that is not capable of matching the data exactly.
Retrotransposon	Virus-like stretches of DNA that can amplify and insert themselves back into the genome. Retrotransposons are very frequent in plants, but also common in higher animals.
Reverse engineering	A conglomerate of mathematical, statistical, and computational methods for inferring the structure and/or internal mechanisms of a system from observations of outward responses of the system.
Ribosome	A molecular machine, consisting of proteins and specific RNAs, that translates the genetic information encoded in an mRNA into the corresponding protein.
Robustness	The somewhat ill-defined concept of the tolerance of a system or model toward perturbations. Robustness is often partially assessed with sensitivity analyses.
Schwann cell	Cells of the peripheral nervous system that electrically insulate the axons of neurons by wrapping sheaths of lipid materials around them. They also aid in the fast transmission of nerve impulses.
Search algorithm	Any of numerous computational optimization methods that attempt to find one or more optimal solutions to a complex problem. It uses many iterations of some cleverly guided trial-and-error scheme that successively improves the currently best solution. An example is a genetic algorithm.
Self-organization	The appearance of patterns out of an unordered state, due to rules governing the

	activities and interactions among the components of a system.
Sensitivity analysis	A core diagnostic method of systems analysis that assesses how much a feature like the steady state of a system is affected if a parameter value is slightly changed.
Sensory neuron	A nerve cell that is activated by external information, such as touch, sound, smell, and light, and transmits this information electrically to other nerve cells.
Signal	A molecule or chemical or physical event that transmits information from the environment, a cell or an organism to another cell or organism. In some cases a cell can even send a signal to itself and respond to it.
Signal transduction (signaling)	Any means of transmitting a signal from a sender to a receiver. Often, both the senders and receivers are cells.
Signaling cascade	Key component of many signal transduction systems inside a cell. A typical cascade consists of three layers of proteins that can be either phosphorylated or dephosphorylated. In the phosphorylated form, the protein facilitates the phosphorylation at the next layer. Signaling cascades integrate and amplify incoming signals and filter out noise.
Silencing	The suppression of gene expression by certain small RNAs.
Simulation	Key method of systems biology, in which a model is repeatedly analyzed with different parameter settings or inputs. Some simulations explore specific "What-if" scenarios, whereas large-scale Monte Carlo simulations are used to establish the repertoire of possible responses of a model and identify best-case, worst-case, and most likely scenarios.
Single nucleotide polymorphism (SNP)	Genetic mutation that alters one nucleotide. Most SNPs are inconsequential, but

collectively they are major drivers of evolution. Two humans typically differ by thousands of SNPs. Pronounced "snip."

Sloppiness (modeling)
Observation that all parameter settings of a model within a certain domain can fit the same data similarly well.

Small RNA
One of a class of short RNAs that can directly affect the function of mRNAs or form regulatory complexes with some proteins.

Small-world property (of networks)
Often encountered composition of networks where a few nodes, called hubs, have numerous connections to otherwise only sparsely connected nodes. Advantages are that the average number of steps required to move from any one node to any other node is minimal, and that failure of any of the sparsely connected nodes usually only causes limited damage. An example is the airport system.

Sphingolipid
A compound within a specific class of lipids that is an important component of cell membranes and can also have signaling functions.

State (of a system or model)
A vague term, which, however, can be crisply defined for a given situation, to represent a collection of features of a system or model at a given point in time.

Static (model)
Not changing over time.

Steady state
Often the normal state of a system or model where none of the variables changes in magnitude, although material is being exchanged among the variables. Biologists sometimes refer to this state as homeostasis.

Stem cell
A cell that is not fully differentiated and gives rise to daughter cells, one of which is again a stem cell, while the other daughter is a differentiated cell or a different stem cell. The stem cell with the greatest variety of ultimate offspring is a fertilized egg cell. The body contains numerous stem cells that may

differentiate into alternate cell types. Adult stem cells include cells of the blood forming system and the lining of the gut.

Stimulus
A change in external conditions or some other perturbation that is expected to lead to a system response. An extended series of stimuli is a widely used simulation tool of systems analysis.

Stochastic
Containing unpredictable random aspects, such as the result of a lottery.

Stress
An external or internal situation that alters the normal physiological state of a cell or organism and requires a system response for returning to normalcy or for at least tolerating the situation.

Superposition principle
A fundamental characteristic of linear, but not of nonlinear, systems: namely, the sum of the responses of a system to two individual stimuli is the same as the single system response if the two stimuli are applied simultaneously. Synergism and antagonism violate this principle and are found only in nonlinear systems.

Support vector machine
A specific, supervised learning algorithm within the field of machine learning, used for classification tasks, such as distinguishing normal cells from cancer cells.

Synapse
A functional connection between the axon of a neuron and the dendrite of another neuron that is the base unit for transducing nerve impulses through neurotransmitters.

Synergism
As already observed by Aristotle, the phenomenon that "the sum can be greater than the parts." In contrast to superposition, two simultaneous inputs lead to a stronger response than the sum of responses to the two stimuli when applied separately.

Synthetic biology	A new branch of biology that attempts to create new or modified organisms based on rational design and prediction. In some sense, an application of systems biology.
System	A well-structured assembly of components that interact with each other in a dynamic fashion.
Systems biology	A biological field of study that uses computational and experimental methods to investigate biological phenomena comprehensively and within their natural contexts.
Tagging (molecular biology)	The marking of a type of molecule, such as a protein, with an artificial marker that can reveal the location of the molecule.
Taxonomy	A sophisticated classification scheme exhibiting the relatedness among species or other entities. The most prominent example in biology is Carl Linnaeus' (1735) classification of plants and animals.
T-cell	T-cell lymphocyte. This class of white blood cells comes in several varieties, including $CD4^+$ T helper cells, $CD8^+$ cytotoxic T-cells, and regulatory suppressor T-cells.
Time series	Datasets consisting of measurements at several or many subsequent time points.
Toggle switch (genetics)	A simplified phenomenon or its representation whereby a gene is either entirely silent or maximally expressed.
Topology	The study of general spatial relationships that are unaffected by gradual distortions.
Training data	Data used to parameterize a model. An example is a large dataset that allows a machine learning algorithm to calibrate its parameters in order to distinguish between healthy and cancer cells.
Transcription	The process of creating mRNA that is complementary to a DNA sequence.

Transcription factor	A protein that can bind to a DNA sequence and by doing so affects the expression of a gene.
Translation	The synthesis of a protein from amino acids according to the genetic code represented in an mRNA. Ribosomes are the molecular machines executing this translation.
Transposon	A DNA sequence that can be inserted into the DNA of a host organism and move from one location in the genome to another. Also called a transposable element, often originating from a virus.
Ubiquitin	A widely used, small protein that a cell attaches to other proteins and marks them for disassembly by the proteasome.
Uncertainty	An inaccuracy in data due to imperfect measurements. In contrast to variability, which is intrinsic to all organisms and cannot be eliminated, improved measurement methods can in principle reduce uncertainty to almost zero.
Validation	A collection of tests trying to confirm (or refute) the correctness of a model. A typical validation is achieved by successfully fitting the model to data that had not been involved in the model design. A validation is almost never complete.
Variability	A collective term for intrinsic differences between individual cells or organisms. In contrast to uncertainty, which is principally reducible with improved methods, variability cannot be avoided, except through biological strategies such as using organisms that are clones of each other.
Variable	The mathematical representative of a system component of interest.
Vector	1. Biology: A virus, microbe or higher organism used to transport genetic material into

a host. 2. Physics: An entity with a direction and a length. 3. Math: The arrangement of numbers or variables as a one-dimensional array. Used extensively in linear algebra and calculus.

Vesicle   A small compartment, typically within a cell, consisting of a lipid membrane and containing a fluid or metabolite, such as an enzyme, a neurotransmitter or a compound like a hormone that is to be secreted.

Selected further reading

The following general sources might be of interest.

Easy-to-read articles highlighting fundamental issues of systems biology

M.A. Savageau (1991) The challenge of reconstruction, *New Biol.* 3, 101–102.
Y. Lazebnik (2002) Can a biologist fix a radio? Or, what I learned while studying apoptosis, *Cancer Cell* 2, 179–182.

Introductory textbook with many references to materials presented here

E.O. Voit (2012) *A First Course in Systems Biology,* Garland Science, New York and London.

Books with little or no math

A.-L. Barabási (2003) *Linked: How Everything is Connected to Everything Else and What it Means for Business, Science, and Everyday Life,* Penguin, Harmondsworth, UK.
M. Newman, A.-L. Barabási, and D.J. Watts (Eds.) (2006) *The Structure and Dynamics of Networks,* Princeton University Press, Princeton, NJ.
D. Noble (2006) *The Music of Life: Biology Beyond Genes,* Oxford University Press, Oxford, UK.

Introductory texts with some math

U. Alon (2006) *An Introduction to Systems Biology: Design Principles and Biological Circuits,* Chapman & Hall/CRC, London, UK.
E. Klipp, W. Liebermeister, C. Wierling, H. Lehrach, and R. Herwig (2009) *Systems Biology,* Wiley-Blackwell, Weinheim, Germany.

Texts with a good dose of math

A. Kremling (2014) *Systems Biology: Mathematical Modeling and Model Analysis*, Chapman & Hall, Abingdon, UK.

B.Ø. Palsson (2015) *Systems Biology: Constraint-based Reconstruction and Analysis*, Cambridge University Press, Cambridge, UK.

E.O. Voit (2000) *Computational Analysis of Biochemical Systems: A Practical Guide for Biochemists and Molecular Biologists*, Cambridge University Press, Cambridge, UK.

Looking for hard numbers in biology?

BioNUMB3R5: The Database of Useful Biological Numbers, http://bionumbers .hms.harvard.edu/

Specific sources

Chapter 1

M.A. Savageau (1991) The challenge of reconstruction, *New Biol.* 3, 101–102.

M.A. Savageau (1991) Reconstructionist molecular biology, *New Biol.* 3, 190–197.

Chapter 2

E. Gubb, and R. Matthiesen (2010) Introduction to omics, *Methods in Mol. Biol.* 593, 1–23.

Chapter 3

P. Flach (2012) *Machine Learning: The Art and Science of Algorithms that Make Sense of Data*, Cambridge University Press, Cambridge, UK.

P. Simon (2013) *Too BIG to IGNORE*, John Wiley & Sons, Hoboken, NJ.

A.M. Turing (1950) Computing machinery and intelligence, *Mind* 59, 433–460.

Chapter 4

R. Alves, and A. Sorribas (2011) Special issue on biological design principles, *Math Biosci.* 231(1).

G. Galilei (1638) *Dialogues Concerning Two New Sciences*, Macmillan Company, NY 1914; reprinted by Dover Books on Physics, 1954.

Chapter 5

A. Ben-Yakar, N. Chronis, and H. Lu (2009) Microfluidics for the analysis of behavior, nerve regeneration, and neural cell biology in *C. elegans*, *Curr. Opin. Neurobiol.* 19, 561–567.

B.-C. Chen, W.R. Legant, K. Wang, et al. (2014) Lattice light-sheet microscopy: Imaging molecules to embryos at high spatiotemporal resolution, *Science* 346. DOI: 10.1126/science.1257998.

L.R. Girard, T.J. Fiedler, T.W. Harris, et al. (2007) WormBook: The online review of *Caenorhabditis elegans* biology, *Nucleic Acids Res.* 35, D472–D475.

M.C. Leung, P.L. Williams, A. Benedetto, C. Au, et al. (2008) *Caenorhabditis elegans*: An emerging model in biomedical and environmental toxicology, *Toxicol. Sci.* 106, 5–28.

D. Mark, S. Haeberle, G. Roth, F. von Stetten, and R. Zengerle (2010) Microfluidic lab-on-a-chip platforms: Requirements, characteristics and applications, *Chem. Soc. Rev.* 39, 1153–1182.

L.F. Marvin, M.A. Roberts, and L.B. Fay (2003) Matrix-assisted laser desorption/ ionization time-of-flight mass spectrometry in clinical chemistry, *Clin. Chim. Acta* 337, 11–21.

J.L. Norris, and R.M. Caprioli (2013) Analysis of tissue specimens by matrix-assisted laser desorption/ionization imaging mass spectrometry in biological and clinical research, *Chem. Rev.* 113, 2309–2342.

S.Y. Rojahn (2014) Does Illumina have the first $1,000 genome? *Biomedicine News*, www.technologyreview.com/news/523601/does-illumina-have-the-fi rst-1000-genome/.

Chapter 6

V. Bush (1936) Instrumental analysis, *Bull. Amer. Math. Soc.* 42 (10), 649–669.

A.T. Johnson (2013) Teaching the principle of biological optimization, *J. Biol. Eng.* 7, 6.

Chapter 7

M.V. Henri, (1903) *Lois générales de l'action des diastases*, Hermann, Paris.

O.W. Holmes (1858 [2006]) *The Autocrat of the Breakfast Table*, Echo Library, Teddington, UK.

L. Michaelis, and M.L. Menten (1913) Die Kinetik der Invertinwirkung. *Biochemische Zeitschrift* 49, 333–369.

N.V. Torres, and E.O. Voit (2002) *Pathway Analysis and Optimization in Metabolic Engineering*, Cambridge University Press, Cambridge, UK, Chapters 3 and 6.

E.O. Voit, H.A. Martens, and S.W. Omholt (2015) 150 years of the mass action law, *PLoS Comp. Biol.* 11 (1), e1004012.

Chapter 8

E.O. Voit (2008) Modeling metabolic networks using power-laws and S-systems, *Essays in Biochemistry* 45, 29–40.

Chapter 9

M.A. Bedau (1997) Weak emergence, in J. Tomberlin (Ed.) *Philosophical Perspectives: Mind, Causation, and World*, Vol. 11, Blackwell, Malden, MA, 375–399, people.reed.edu/~mab/papers/weak.emergence.pdf.

D. Dörner (1996) *The Logic of Failure: Recognizing and Avoiding Error in Complex Situations*, Perseus Books, Cambridge, MA.

J.B.S. Haldane (1932) *The Causes of Evolution*, Longmans, Green and Co., London, UK.

A.K. Konopka (2007) *Systems Biology: Principles, Methods, and Concepts*, CRC Press, Boca Raton, FL.

E. Mayr (2004) *What Makes Biology Unique?* Cambridge University Press, New York, NY.

T. O'Connor (2012) Emergent properties, *Stanford Encyclopedia of Philosophy*, plato.stanford.edu/entries/properties-emergent/.

Chapter 10

A.L. Goldberger (1990) Is the normal heartbeat chaotic or homeostatic?, *Physiology* 6 (2), 87–91.

L. Liebovitch (1998) *Fractals and Chaos*, Oxford University Press, New York, NY.

L.A. Lipsitz (2004) Physiological complexity, aging, and the path to frailty, *Sci. Aging Knowl. Environ.* DOI: 10.1126/sageke.2004.16.pe16.

Chapter 11

L.H. Epstein, J.L.Temple, B.J. Neaderhiser, R.J. Salis, et al. (2007) Food reinforcement, the dopamine D2 receptor genotype, and energy intake in obese and nonobese humans, *Behav. Neurosci.* 121 (5), 877–886.

K.M. Eyster (1998) Introduction to signal transduction: A primer for untangling the web of intracellular messengers, *Biochem. Pharm.* 55 (12), 1927–1938.

M. Kojima, and K. Kangawa (2005) Ghrelin: Structure and function, *Physiol. Rev.* 85 (2), 495–522.

Chapter 12

A.-L. Barabási, and Z. Oltvai (2004) Network biology: Understanding the cell's functional organization, *Nat. Rev. Gen.* 5, 101–113.

A. Ma'ayan (2011) Introduction to network analysis in systems biology, *Sci. Signal* 4 (190), tr5.

G.A. Pavlopoulos, M. Secrier, C.N. Moschopoulos, T.G. Soldatos, et al. (2011) Using graph theory to analyze biological networks, *BioData Mining* 4, 10.

Chapter 13

J.K. Bowmaker (1998) Evolution of colour vision in vertebrates. *Eye* 12 (3b), 541–547.

N. Eldredge, and S. J. Gould (1972) Punctuated equilibria: An alternative to phyletic gradualism, in T.J.M. Schopf (Ed.) *Models in Paleobiology*, Freeman Cooper, San Francisco, CA, 82–115. Reprinted in N. Eldredge (1985) *Time Frames*, Princeton University Press, Princeton, NJ, 193–223.

L. Loewe (2012) How evolutionary systems biology will help understand adaptive landscapes and distributions of mutational effects, *Adv. Exp. Med. Biol.* 751, 399–410.

M. Lynch (2010) Evolution of the mutation rate, *Trends Genet.* 26 (8), 345–352.

K.L. Spalding, R.D. Bhardwaj, B.A. Buchholz, H. Druid, and J. Frisén (2005) Retrospective birth dating of cells in humans, *Cell* 122 (1), 133–143.

P. Tamayo, D. Slonim, J. Mesirov, Q. Zhu, et al. (1999) Interpreting patterns of gene expression with self-organizing maps: Methods and application to hematopoietic differentiation. *Proc. Nat. Acad. Sci. USA* 96 (6), 2907–2912.

J.J. Tyson, and B. Novak (2008) Temporal organization of the cell cycle, *Curr. Biol.* 18 (17), R759–R768.

Chapter 14

M.R. Rondon, P.R. August, A.D. Bettermann, S.F. Brady, et al. (2000) Cloning the soil metagenome: A strategy for accessing the genetic and functional diversity of uncultured microorganisms, *Appl. Environ. Microbiol.* 66, 2541–2547.

J.T. Trevors (2010) One gram of soil: A microbial biochemical gene library, *Antonie Van Leeuwenhoek* 97, 99–106.

F. Turroni, A. Ribbera, E. Foroni, D. van Sinderen, and M. Ventura (2008) Human gut microbiota and bifidobacteria: From composition to functionality, *Antonie Van Leeuwenhoek* 94, 35–50.

Chapter 15

J.J. Barr, R. Auro, M. Furlan, K.L. Whiteson, et al. (2013) Bacteriophage adhering to mucus provide a non-host-derived immunity, *Proc. Nat. Acad. Sci. USA* 110 (26), 10771–10776.

C. Erridge (2013) *Undergraduate Immunology*, Amazon Digital Services, Inc.

E.T. Harvill (2013) Cultivating our "frienemies": Viewing immunity as microbiome management, *mBio* 4 (2), e00027–13.

L.M. Sompayrac (2012) *How the Immune System Works*, Wiley-Blackwell, Hoboken, NJ.

K. Todar (2012) *Todar's Online Textbook of Bacteriology.* http://textbookofbacteriology.net/

Chapter 16

R.L. Buckner, and F.M. Krienen (2013) Association cortices: The evolution of distributed association networks in the human brain, *Trends in Cognitive Sciences* 17 (12), 648–665.

H. Hama, H. Kurokawa, H. Kawano, R. Ando, et al. (2011) Scale: A chemical approach for fluorescence imaging and reconstruction of transparent mouse brain, *Nat. Neurosci.* 14 (11), 1481–1488.

N. Le Novère (Ed.) (2012) *Computational Systems Neurobiology*, Springer Verlag, Dordrecht, The Netherlands.

E. De Schutter E (2008) Why are computational neuroscience and systems biology so separate?, *PLoS Comput. Biol.* 4 (5), e1000078.

http://discovermagazine.com/2011/mar/10-numbers-the-nervous-system#.UvwD6mJdUuk

http://legacy.mos.org/quest/mummy.php

http://medcitynews.com/2013/04/why-obamas-brain-mapping-project-matters/#ixzz2ZzZxZVbK

http://science.howstuffworks.com/mummy2.htm

http://www.sfn.org/awards-and-funding/global-funding-sources

Chapter 17

E.O. Voit, and K.L. Brigham (2008) The role of systems biology in predictive health and personalized medicine, *The Open Path. J.* 2, 68–70.

http://www.mayoclinic.com/health/symptom-checker/DS00671

http://symptoms.webmd.com/#introView

Chapter 18

A. Arkin (2008) Setting the standard in synthetic biology, *Nat. Biotechn.* 26, 771–774.

D.A. Drubin, J.C. Way, and P.A. Silver (2007) Designing biological systems, *Genes. Dev.* 21, 242–254.

A.E. Friedland, T.K. Lu, X. Wang, D. Shi, et al. (2009) Synthetic gene networks that count, *Science* 324, 1199–1202.

J.D. Keasling (2008) Synthetic biology for synthetic chemistry, *ACS Chem. Biol.* 3, 64–76.

W.D. Marner, II (2009) Practical application of synthetic biology principles, *Biotechnol. J.* 4, 1406–1419.

S. Mukherji, and A. van Oudenaarden (2009) Synthetic biology: Understanding biological design from synthetic circuits, *Nat. Rev. Genetics* 10, 859–871.

H. Salis, A. Tamsir, and C. Voigt (2009) Engineering bacterial signals and sensors, *Contrib. Microbiol.* 16, 194–225.

J.J. Tabor, H.M. Salis, Z.B. Simpson, A.A. Chevalier, et al. (2009) A synthetic genetic edge detection program, *Cell* 137, 1272–1281.

C.T. Trinh, P. Unrean, and F. Srienc (2008) Minimal *Escherichia coli* cell for the most efficient production of ethanol from hexoses and pentoses, *Appl. Env. Microbiol.* 74, 3634–3643.

Chapter 19

P.L. Galison (1997) *Image and Logic: A Material Culture of Microphysics*, University of Chicago Press, Chicago, IL.

National Research Council (2003) *Beyond Productivity: Information, Technology, Innovation, and Creativity*, National Academies Press, Washington, DC.

National Research Council (2009) *A New Biology for the 21st Century*, National Academies Press, Washington, DC.

R.J. Spiro (1988) *Cognitive Flexibility Theory: Advanced Knowledge Structures in Ill-structured Domains*, University of Illinois Press, Champaign, IL.

S.L. Star, and J.R. Griesemer (1989) Institutional ecology, "translations" and boundary objects: Amateurs and professionals in Berkeley's Museum of Vertebrate Zoology, *Social Studies of Science* 19, 387–420.

Index